SpringerBriefs in Mathematics

SpringerBriefs in Mathematics showcases expositions in all areas of mathematics and applied mathematics. Manuscripts presenting new results or a single new result in a classical field, new field, or an emerging topic, applications, or bridges between new results and already published works, are encouraged. The series is intended for mathematicians and applied mathematicians. All works are peer-reviewed to meet the highest standards of scientific literature.

BCAM SpringerBriefs

BCAM *SpringerBriefs* aims to publish contributions in the following disciplines: Applied Mathematics, Finance, Statistics and Computer Science. BCAM has appointed an Editorial Board, who evaluate and review proposals.

Typical topics include: a timely report of state-of-the-art analytical techniques, bridge between new research results published in journal articles and a contextual literature review, a snapshot of a hot or emerging topic, a presentation of core concepts that students must understand in order to make independent contributions.

Please submit your proposal to the Editorial Board or to Francesca Bonadei, Executive Editor Mathematics, Statistics, and Engineering: francesca.bonadei@springer.com.

More information about this series at http://www.springer.com/series/10030

Genni Fragnelli · Dimitri Mugnai

Control of Degenerate and Singular Parabolic Equations

Carleman Estimates and Observability

Genni Fragnelli
Department of Mathematics
University of Bari Aldo Moro
Bari, Italy

Dimitri Mugnai
Department of Biological and Ecological
Sciences
Tuscia University
Viterbo, Italy

ISSN 2191-8198   ISSN 2191-8201   (electronic)
SpringerBriefs in Mathematics
ISBN 978-3-030-69348-0   ISBN 978-3-030-69349-7   (eBook)
https://doi.org/10.1007/978-3-030-69349-7

This Springer imprint is published by the registered company Springer Nature Switzerland AG
The registered company address is: Gewerbestrasse 11, 6330 Cham, Switzerland

With infinite love to Neri, Artemisia and Manfredi

Preface

Abbrivo

Controllability issues for parabolic problems have been a mainstream topic in recent years, and several developments have been pursued: starting from the heat equation in bounded and unbounded domain, related contributions have been found for more general situations. A common strategy in showing controllability results is to prove that certain global Carleman estimates hold true for the operator which is the adjoint of the given one, and, from them, to find related observability inequalities for the solution of the initial problem.

In these notes we follow this approach focusing on some classes of parabolic operators of the form

$$u_t - \operatorname{div}(a(x)\nabla u) - \frac{\lambda}{b(x)}u, \tag{1}$$

or

$$u_t - a(x)\Delta u - \frac{\lambda}{b(x)}u, \tag{2}$$

associated to homogeneous Dirichlet boundary conditions. Here $(t, x) \in Q := (0, T) \times \Omega$, $T > 0$ being a fixed number and Ω a bounded smooth domain of $\mathbb{R}^N$, $N \geq 1$, and $\lambda \in \mathbb{R}$.

As for a and b, they can be strictly positive or can degenerate somewhere: if $a = 0$ somewhere in Ω, the problem becomes *degenerate*, while if $b = 0$, it is *singular*. Of course, the two occurrences can happen simultaneously, and the problem is *degenerate and singular*.

If $\Omega = (0, 1)$, the prototypes we have in mind are $a(x) = x^{\alpha_1}(1 - x)^{\alpha_2}$ and $b(x) = x^{K_1}(1 - x)^{K_2}$, if the degeneracy occurs at the boundary of the domain, $a(x) = |x - x_0|^{K_1}$ and $b(x) = |x - x_0|^{K_2}$, if it occurs in an interior point $x_0 \in (0, 1)$, for some $\alpha_i, K_i \geq 0$, $i = 1, 2$; notice that, if $\alpha_1 > 0$ and $\alpha_2 = 0$ we have degeneracy only on one side of the boundary, and also such situations can be arranged. We consider the case in which a and b degenerate at the same point, since this is the most difficult situation: indeed, if the degeneracy and the singularity take place in different points, one can treat them separately.

Per l'alto mare aperto[1]

Carleman estimates for uniformly parabolic operators without degeneracies or singularities have been largely developed (see, e.g., Fursikov–Imanuvilov [87], one of the milestones on this topic). However, recently, these estimates have been also studied for operators which are not uniformly parabolic. Indeed, as pointed out by several authors, many problems coming from Physics (see [94]), Biology (see [52]), and Mathematical Finance (see [91]) are described by parabolic equations which admit some kind of degeneracy. As a consequence, in the last recent years an increasing interest has been devoted to (1) or (2) in the case $\lambda = 0$, see [1, 9, 18–20, 28, 30, 34, 36, 37, 70, 81–85], immediately followed by studies on the case $\lambda \neq 0$, see [68, 69, 71, 79, 80, 83, 92, 107] and [115], just to mention a few.

Though here we focus only on linear equations, it is mandatory to recall that nonlinear cases where studied, as well; for instance, see [50, 93, 120] and [123].

In order to clarify our subject, let us consider a general problem

$$\begin{cases} u_t - Au - \frac{\lambda}{b(x)}u = f(t, x)\chi_\omega, & (t, x) \in Q, \\ u(0, x) = u_0(x) \in X, & x \in \Omega, \end{cases}$$

with suitable boundary conditions, where χ_ω is the characteristic function of the set $\omega \subseteq \overline{\Omega}$, X is a suitable Hilbert space where the problem is settled and A is the underlying operator; of course, for the models (1) and (2) we have $Au = \text{div}(a(x)\nabla u)$ and $Au = a(x)\Delta u$, respectively. We recall that such a problem is said to be *globally null controllable*—(GNC) for short—if for every $u_0 \in X$ there exists $f \in L^2(0, T; X)$ such that

$$u(T, x) = 0 \text{ for every } x \in \Omega$$

and

$$\|f\|_{L^2(0,T;X)} \leq C\|u_0\|_X$$

for some universal positive constant C. In order to arrive at such a result, one way is to prove an *observability inequality*, i.e., an estimate of the form

$$\int_\Omega v^2(0, x)dx \leq C \int_0^T \int_\omega v^2(t, x)dxdt, \tag{3}$$

---

[1]Dante Alighieri, *La Divina Commedia, Inferno*, C. XXVI, v. 100.

for the solutions of the adjoint problem. On the other hand, inequalities of this type are a consequence of *global Carleman estimates* of the form

$$\int_Q \{sw_1(t,x)v^2(t,x) + s^3 w_2(t,x)v_x^2(t,x)\}e^{2s\phi}dxdt \le C\int_Q e^{2s\phi}f^2 dxdt, \quad (4)$$

where $C > 0$ is a universal constant, w_1, w_2 are two weights (possibly degenerate or singular), $s \ge s_0$ where s_0 is a positive parameter and ϕ is a suitable weight. Inequalities of this form are named after Carleman's paper [40], where he obtained such inequalities in order to show a unique continuation property for solutions of certain parabolic problems. In Chap. 1, we will see how controllability follows from observability, while the derivation of observability from Carleman estimates is, in general, more complicated and depends on the different treated cases. In any case, the main purpose of this volume is to show how obtaining inequalities like (4) and then some like (3).

The inspiring situation is the case of parabolic operators with singular inverse-square potentials. First results in this direction were obtained in [117] for the *non degenerate* singular potentials with the Laplacian operator

$$u_t - \Delta u - \lambda \frac{1}{|x|^2}u, \ (t,x) \in Q, \quad (5)$$

associated to Dirichlet boundary conditions in a bounded domain $\Omega \subset \mathbb{R}^N$ containing the singularity $x = 0$ in the interior (see also [116] for the wave and Schrödinger equations and [41] for boundary singularity). Similar operators of the form

$$u_t - \Delta u - \lambda \frac{1}{|x|^K}u, \ (t,x) \in Q,$$

arise for example in quantum mechanics (see, for example, [5, 47]), or in combustion problems (see, for example, [8, 23, 48, 88]), and it is known that they generate interesting phenomena. For example, in [5] and in [6] it was proved that, for all values of λ, global positive solutions exist if $K < 2$, whereas instantaneous and complete blow-up occurs if $K > 2$. In the critical case, i.e., $K = 2$, the value of the parameter λ determines the behavior of the equation: if $\lambda \le 1/4$ (which is the optimal constant of the Hardy inequality, see [22]) global positive solutions exist, while, if $\lambda > 1/4$, instantaneous and complete blow-up occurs (for other comments on this argument we refer to [115]). We recall that in [117], Carleman estimates were established for (5) under the condition $\lambda \le 1/4$. On the contrary, if $\lambda > 1/4$, in [53] it was proved that null controllability may fail, unless special initial data are considered, see [117, Theorem 5.1].

On the other hand, purely degenerate operators like

$$u_t - (a(x)u_x)_x,$$

were firstly attacked by using Carleman estimates in [34] and [36] for $a(x) = x^\alpha$, $\alpha \in$ (0, 2). Lately, more general functions $a(x) \sim x^\alpha$, $\alpha \in (0, 2)$, have been considered starting with [1].

We remark that the non-degenerate singular problems studied in [5, 41, 53, 115–117] cover the multidimensional case, while here we treat essentially the case $N = 1$, as Vancostenoble in [115], who studied the operator that couples a degenerate diffusion coefficient with a singular potential. In particular, for $\alpha \in [0, 2)$ and $K \leq 2 - \alpha$, the author established Carleman estimates for the operator

$$u_t - (x^\alpha u_x)_x - \lambda \frac{1}{x^K} u, \ (t, x) \in Q,$$

unifying the results of [36] and [117] in the purely degenerate operator and in the purely singular one, respectively. This result was then extended in [68] and in [69] to the operators

$$u_t - (a(x)u_x)_x - \lambda \frac{1}{x^K} u, \ (t, x) \in Q, \tag{6}$$

for $a \sim x^\alpha$, $\alpha \in [0, 2)$ and $K \leq 2 - \alpha$. Here, as before, the function a degenerates at the boundary of the space domain, and Dirichlet boundary conditions are in force.

We remark the fact that all the papers cited so far, with the exception of [53], consider a singular/degenerate operator with degeneracy or singularity appearing at the boundary of the domain. For example, in (6) as a one can consider the double power function

$$a(x) = x^{\alpha_1}(1 - x)^{\alpha_2}, \quad x \in [0, 1],$$

for some $\alpha_i > 0$. To the best of our knowledge, [20, 81] and [82] are the first papers dealing with Carleman estimates (and, consequently, null controllability) for operators (in divergence and in non-divergence form with Dirichlet or Neumann boundary conditions) with mere degeneracy at the interior of the space domain (for related systems of degenerate equations we refer to [19]). We also recall [75] and [76] for other types of control problems associated to parabolic operators with interior degeneracy in divergence and non-divergence form, respectively.

We emphasize the fact that an interior degeneracy does not imply a simple adaptation of previous results and of the techniques used for boundary degeneracy. For instance, imposing homogeneous Dirichlet boundary conditions, in the latter case one knows *a priori* that any function in the reference functional space vanishes exactly at the degeneracy point. On the other hand, when the degeneracy point is in the interior of the spatial domain, such information is not valid anymore, and we cannot take advantage of this fact.

For this reason, part of these notes is devoted to study the operator

$$u_t - (au_x)_x - \frac{\lambda}{b(x)}u$$

or

$$u_t - au_{xx} - \frac{\lambda}{b(x)}u$$

that couples a *general* degenerate diffusion coefficient with a *general* singular potential with degeneracy and singularity at the *interior* of the space domain. In particular, under suitable conditions on all the parameters of the operator, we establish Carleman estimates and, as a consequence, null controllability for the associated generalized heat problem. Clearly, this result generalizes the one obtained in [81] or [82]: in fact, if $\lambda = 0$ (that is, if we consider the purely degenerate case), we recover the main contributions therein. However, we underline that in [82] more irregular coefficients are considered: indeed, though we have in mind prototypes as power functions for the degeneracy and the singularity, in [82] we do not limit our investigation to these functions, which are analytic out of their zero.

The classical approach to study singular operators in dimension 1 relies in the validity of the Hardy–Poincaré inequality

$$\int_0^1 \frac{u^2}{x^2}dx \le 4 \int_0^1 (u')^2 dx, \tag{7}$$

which is valid for every $u \in H^1(0, 1)$ with $u(0) = 0$. Similar inequalities are the starting point to prove well posedness of the associated problems in the Sobolev spaces under consideration.

If the degeneracy and the singularity occur in the interior of the space domain we give an inequality related to (7), but with a degeneracy coefficient in the gradient term; such an estimate is valid in a suitable Hilbert space $\mathcal{H}$ we shall introduce below, and it states the existence of $C > 0$ such that for all $u \in \mathcal{H}$ we have

$$\int_0^1 \frac{u^2}{b}dx \le C \int_0^1 a(u')^2 dx.$$

This inequality, which is related to another weighted Hardy–Poincaré inequality (see Proposition 1.11), is the key step for the well posedness of the considered problem. Once this is done, global Carleman estimates follow, provided that a suitable choice of the weight functions is made.

We remark that degenerate problems have been faced essentially only in the one-dimensional case, except for a special situation treated in [35] for $N = 2$. No surprise, then, if the degenerate/singular problems we shall treat are confined to the case $N = 1$.

The First Thing to Realize About Parallel Universes is that they are not Parallel[2]

Before concluding, it is worth mentioning that controllability is a mainstream topic since many years and that other approaches are available. Here we briefly comment on some parallel ones.

The *moment method* was originally introduced in [61] (and lately developed by several other authors, for instance see [108] and [113]). In order to shortly describe it, which inspired the approaches in Chap. 3, consider the model problem

$$\begin{cases} u_t - u_{xx} = f(x)y(t), & (t, x) \in (0, T) \times (0, \pi), \\ u(t, 0) = u(t, 1) = 0, \ t \in (0, t), \\ u(0, x) = u_0(x), & x \in (0, \pi), \end{cases}$$

where $f, u_0 \in L^2(0, \pi)$ and one wants a control $y \in L^2(0, T)$ such that $u(T, x) = 0$ for all $x \in (0, \pi)$. Now, choose $e_k(x) = \sqrt{\frac{2}{\pi}} \sin(kx)$, $k \geq 1$, as the k-th element of an orthonormal basis in $L_2(0, \pi)$ with associated eigenvalue $\lambda_k = k^2$. Then, look for a solution of the form

$$u(t, x) = \sum_{k=1}^{\infty} u_k(t)e_k(x).$$

If

$$f(x) = \sum_{k=1}^{\infty} f_k e_k(x) \text{ and } u_0(x) = \sum_{k=1}^{\infty} u_k^0 e_k(x),$$

then such a u exists if and only if there exists $y \in L^2(0, T)$ such that (see [61])

$$-e^{-k^2 T} u_k^0 = f_k \int_0^T ke^{-k^2(T-t)} y(t)dt \text{ for all } k \geq 1,$$

a classical *moment problem* in $L^2(0, T)$.

Without dwelling on other methods, we simply quote some of them.

The *flatness method* was introduced in [100] and the idea is to look for solutions of parabolic problems in terms of series different from those in Fourier-style and using Gevrey classes.

The *return method* was introduced in [43] and its idea is to reduce null controllability (and not only) of a nonlinear system to the controllability of linearized systems.

---

[2]Douglas Adams, *Mostly Harmless, The Hitchhiker's Guide to the Galaxy.*

The *transmutation method* was introduced in [103] and it shows that the exact controllability of some hyperbolic equations at time T implies the null controllability of a related parabolic equation at any time.

On a Circle Every Starting Point Can Also be an End Point[3]

This volume is intended for graduated students who wish to approach this subject and to senior researchers who can find many aspects in one single text. In Chap. 1 we collect some essential Hardy-type inequalities and we will prove the equivalence between null controllability and the validity of an observability inequality, which is a key tool in the strategy used in many quoted references. In Chap. 2 we give some controllability results for the degenerate case and in Chap. 3 those for a singular equation. In Chap. 4 we study the case of an equation which is simultaneously degenerate and singular at the boundary of the spatial domain, while in Chap. 5 we consider the case of an interior degenerate/singular equation.

We conclude with two comments. First, we shall consider only problems in presence of homogeneous Dirichlet boundary conditions, the Neumann case being treated in other papers (for instance in [20, 70, 77, 79]). Related results are available for Robin boundary conditions (for example, see [15, 27, 63, 64]).

Second, we shall not report all the proofs of the stated results: we have chosen only some proofs, with a sure lack of completeness, but with a more fluent reading (we hope); of course, we could have made different selections, but we are confident that our merciful "twenty-five readers"[4] will be indulgent towards our personal choices.

Bari, Italy Genni Fragnelli
Viterbo, Italy Dimitri Mugnai
February 2020

---

[3] Heraclitus.
[4] A. Manzoni, "I Promessi sposi" ("The Betrothed").

Acknowledgements

The authors thank the FFABR "Fondo per il finanziamento delle attività base di ricerca" 2017. G. F. is also supported by the Italian MIUR project *Qualitative and quantitative aspects of nonlinear PDEs* (2017JPCAPN), the INdAM-GNAMPA Project 2019 *Controllabilità di PDE in modelli fisici e in scienze della vita* and by Fondi di Ateneo 2017–2018 *Problemi differenziali non lineari*. D. M. is also supported by the Italian MIUR project *Variational methods, with applications to problems in mathematical physics and geometry* (2015KB9WPT_009) and by the INdAM-GNAMPA Project 2020 *Equazioni alle derivate parziali: problemi e modelli*.

Contents

Acronyms

GNC   Globally Null Controllable/Global Null Controllability
SD   Strongly Degenerate
SSD   Strongly-Strongly Degenerate
SWD   Strongly-Weakly Degenerate
WD   Weakly Degenerate
WSD   Weakly-Strongly Degenerate
WWD   Weakly-Weakly Degenerate
Ω domain in $\mathbb{R}^N$, $N \geq 1$
Q_T $(0, T) \times (0, 1)$
Q $(0, T) \times \Omega$
ψ' derivative of a function ψ depending only on x
$\dot{\Theta}$ derivative of a function Θ depending only on t
u_x^2 $(u_x)^2$

Chapter 1
Mathematical Tools and Preliminary Results

Abstract We give some fundamental definitions and some Hardy-type inequalities with boundary or interior degeneracy. We also show the equivalence between null controllability and observability inequality.

Keywords Weak and strong degeneracy · Hardy inequalities · Null controllability · Observability inequality

In this chapter we will give some important inequalities and properties needed to prove well posedness and null controllability for the degenerate/singular parabolic problems we have in mind. In the second part of the chapter we prove that null controllability is equivalent to the observability for the associated homogeneous adjoint problem.

We start recalling the notions of degeneracy we will treat. Of course, different definitions are possible, but related results are equally expected. Based on [1], we classify the degeneracy of a function in two possible ways:

Definition 1.1 **Weakly degenerate case (WD):** a function g is said to be *weakly degenerate* at $x_0 \in [0, 1]$ if $g \in C[0, 1] \cap C^1([0, 1] \setminus \{x_0\})$, $g(x_0) = 0$, $g > 0$ on $[0, 1] \setminus \{x_0\}$ and there exists $K_g \in (0, 1)$ such that $(x - x_0)g'(x) \leq K_g g(x)$ for all $x \in [0, 1] \setminus \{x_0\}$.

Definition 1.2 **Strongly degenerate case (SD):** a function g is said to be *strongly degenerate* at $x_0 \in [0, 1]$ if $g \in C^1([0, 1] \setminus \{x_0\}) \cap W^{1,\infty}(0, 1)$, $g(x_0) = 0$, $g > 0$ on $[0, 1] \setminus \{x_0\}$ and there exists $K_g \in [1, 2)$ such that $(x - x_0)g'(x) \leq K_g g(x)$ for all $x \in [0, 1] \setminus \{x_0\}$.

Observe that the regularity condition in the latter definition is trivially satisfied when $g \in C^1[0, 1]$ (see [1] for the case $x_0 = 0$).

Remark 1.1 In this volume we present the easy types of degeneracy above. More sophisticated ones, with lack of regularity (and much more complicated computations) can be found in [78, 83].

Remark 1.2 In the definitions above, and from now on in this chapter, we have denoted with ′ the derivative of a function depending only on one variable, while derivatives for functions of several variables will be denoted, as usual, with subscript letters, like u_x, u_t and so on.

First of all, we give the following lemma which is crucial in every situation:

Lemma 1.1 (Lemma 2.1 [77]) *Let $x_0 \in [0, 1]$. Assume that a is* **(WD)** *or* **(SD)**.

1. *Then for all $\gamma \geq K_a$ the map*

$$x \mapsto \frac{|x - x_0|^\gamma}{a} \text{ is nonincreasing on the left of } x = x_0$$

and nondecreasing on the right of $x = x_0$,

$$\text{so that } \lim_{x \to x_0} \frac{|x - x_0|^\gamma}{a(x)} = 0 \text{ for all } \gamma > K_a.$$

2. *If $K_a < 1$, then $\dfrac{1}{a} \in L^1(0, 1)$.*

3. *If $K_a \in [1, 2)$, then $\dfrac{1}{\sqrt{a}} \in L^1(0, 1)$ and $\dfrac{1}{a} \notin L^1(0, 1)$.*

Proof We do the proof if $x_0 \in (0, 1)$, the calculations being easier if $x_0 = 0$ or $x_0 = 1$.

The first point is an easy consequence of the definition of degeneracy, so let us prove the second point. By the first part, we get that

$$\frac{|x - x_0|^{K_a}}{a(x)} \leq \max\left\{ \frac{x_0^{K_a}}{a(0)}, \frac{(1 - x_0)^{K_a}}{a(1)} \right\}.$$

Thus

$$\frac{1}{a(x)} \leq \max\left\{ \frac{x_0^{K_a}}{a(0)}, \frac{(1 - x_0)^{K_a}}{a(1)} \right\} \frac{1}{|x - x_0|^{K_a}}.$$

Since $K_a < 1$, the right-hand side of the last inequality is integrable, and then $\dfrac{1}{a} \in L^1(0, 1)$. Analogously, one obtains the integrability stated in the third point.

On the contrary, the fact that $a \in C^1([0, 1] \setminus \{x_0\}) \cap W^{1,\infty}(0, 1)$ and $\dfrac{1}{\sqrt{a}} \in L^1(0, 1)$ implies that $\dfrac{1}{a} \notin L^1(0, 1)$. Indeed, the assumptions on a imply that $a(x) = \displaystyle\int_{x_0}^x a'(s)ds \leq C|x - x_0|$ for a positive constant C. Thus for all $x \neq x_0$, $\dfrac{1}{a(x)} \geq C\dfrac{1}{|x - x_0|} \notin L^1(0, 1)$. $\qquad\square$

1.1 Some Hardy–Poincaré Inequalities with Boundary Degeneracy

In this section we state some lemmas that play an important role in the rest of these notes. As already announced in the Preface, we treat the 1-dimensional case, and so, without loss of generality, we assume that $\Omega = (0, 1)$. We begin with the following Hardy–Poincaré type inequality with a coefficient degenerating at $x_0 = 0$:

Proposition 1.1 (Proposition 2.1 [1]) *Assume that $a \in C[0, 1]$, $a(0) = 0$ and $a > 0$ on $(0, 1]$.*

1. *If there exists $\theta \in (0, 1)$ such that the function*

$$x \mapsto \frac{a(x)}{x^\theta} \text{ is nonincreasing in neighbourhood of } x = 0, \qquad (1.1)$$

   then there is a constant $C > 0$ such that for any function u, locally absolutely continuous on $(0, 1]$, continuous at 0 and satisfying

$$u(0) = 0, \text{ and } \int_0^1 a(x)|u'(x)|^2 \, dx < +\infty,$$

   the following inequality holds:

$$\int_0^1 \frac{a(x)}{x^2} u^2(x) \, dx \le C \int_0^1 a(x)|u'(x)|^2 \, dx . \qquad (1.2)$$

2. *If there exists $\theta \in (1, 2)$ such that the function*

$$x \mapsto \frac{a(x)}{x^\theta} \text{ is nondecreasing in a neighbourhood of } x = 0, \qquad (1.3)$$

   then there is a constant $C > 0$ such that for any function u, locally absolutely continuous on $(0, 1]$ satisfying

$$u(1) = 0, \text{ and } \int_0^1 a(x)|u'(x)|^2 \, dx < +\infty,$$

   inequality (1.2) holds.

For the analogous result of Proposition 1.1 when a degenerates at 1, we refer to [72, Proposition 3.2].

Proof (*Proof of Proposition 1.1*) <u>Case 1.</u> Fix $\beta \in (\theta, 1)$. Since $u(0) = 0$, we have

$$\int_0^1 \frac{a(x)}{x^2} u^2(x) \, dx = \int_0^1 \frac{a(x)}{x^2} \left(\int_0^x (y^{\beta/2} u'(y)) y^{-\beta/2} \, dy \right)^2 dx.$$

Thus

$$\int_0^1 \frac{a(x)}{x^2} u^2(x)\, dx \le \int_0^1 \frac{a(x)}{x^2} \left(\int_0^x (y^\beta |u'(y)|^2\, dy \int_0^x y^{-\beta}\, dy \right) dx.$$

Hence, we have

$$\int_0^1 \frac{a(x)}{x^2} u^2(x)\, dx \le \frac{1}{1-\beta} \int_0^1 \frac{a(x)}{x^{1+\beta}} \left(\int_0^x (y^\beta |u'(y)|^2\, dy \right) dx.$$

By applying Fubini's Theorem, we have

$$\int_0^1 \frac{a(x)}{x^2} u^2(x)\, dx \le \frac{1}{1-\beta} \int_0^1 y^\beta |u'(y)|^2 \left(\int_y^1 \frac{a(x)}{x^{1+\beta}}\, dx \right) dy. \qquad (1.4)$$

Now, by assumption, there exists $\varepsilon > 0$ such that the mapping

$$x \mapsto \frac{a(x)}{x^\theta} \text{ is nonincreasing on } (0, \varepsilon]. \qquad (1.5)$$

We rewrite

$$\int_0^1 y^\beta |u'(y)|^2 \left(\int_y^1 \frac{a(x)}{x^{1+\beta}}\, dx \right) dy = L_\varepsilon + M_\varepsilon + N_\varepsilon,$$

where

$$L_\varepsilon = \int_0^\varepsilon y^\beta |u'(y)|^2 \left(\int_y^\varepsilon \frac{a(x)}{x^{1+\beta}}\, dx \right) dy,$$

$$M_\varepsilon = \int_0^\varepsilon y^\beta |u'(y)|^2 \left(\int_\varepsilon^1 \frac{a(x)}{x^{1+\beta}}\, dx \right) dy,$$

and

$$N_\varepsilon = \int_\varepsilon^1 y^\beta |u'(y)|^2 \left(\int_y^1 \frac{a(x)}{x^{1+\beta}}\, dx \right) dy.$$

For L_ε, we use (1.5), obtaining

$$\int_y^\varepsilon \frac{a(x)}{x^{1+\beta}}\, dx \le \frac{a(y)}{y^\theta} \int_y^\varepsilon x^{\theta-\beta-1}\, dx \le \frac{1}{\beta-\theta} a(y) y^{-\beta}.$$

Using this last inequality for L_ε, we deduce that

$$L_\varepsilon \le \frac{1}{(\beta-\theta)} \int_0^\varepsilon a(x) |u'(x)|^2\, dx. \qquad (1.6)$$

For M_ε, we have

$$M_\varepsilon \leq \frac{1}{\beta} \int_0^\varepsilon a(y)|u'(y)|^2 \frac{y^\beta}{a(y)} \sup_{[\varepsilon,1]}(a)\varepsilon^{-\beta} \, dy \leq \frac{\sup_{[\varepsilon,1]}(a)}{\beta \inf_{[\varepsilon,1]}(a)} \int_0^\varepsilon a(x)|u'(x)|^2 \, dx.$$

(1.7)

For N_ε, we proceed in a similar way and obtain

$$N_\varepsilon \leq \frac{\sup_{[\varepsilon,1]}(a)}{\beta \inf_{[\varepsilon,1]}(a)} \int_\varepsilon^1 a(x)|u'(x)|^2 \, dx.$$

(1.8)

Using (1.6)–(1.8) in (1.4), it follows

$$\int_0^1 \frac{a(x)}{x^2} u^2(x) \, dx \leq C \int_0^1 a(x)|u'(x)|^2 \, dx,$$

(1.9)

where the constant C depends on a, ε, θ and β.

Case 2. Fix $\beta \in (1, \theta)$. Then

$$\int_0^1 \frac{a(x)}{x^2} u^2(x) \, dx = \int_0^1 \frac{a(x)}{x^2} \left(\int_x^1 (y^{\beta/2} u'(y)) y^{-\beta/2} \, dy \right)^2 dx.$$

Thus

$$\int_0^1 \frac{a(x)}{x^2} u^2(x) \, dx \leq \int_0^1 \frac{a(x)}{x^2} \left(\int_x^1 y^\beta |u'(y)|^2 \, dy \int_x^1 y^{-\beta} \, dy \right) dx.$$

Hence, by using Fubini's Theorem, we have

$$\int_0^1 \frac{a(x)}{x^2} u^2(x) \, dx \leq \frac{1}{\beta - 1} \int_0^1 \frac{a(x)}{x^{1+\beta}} \left(\int_x^1 y^\beta |u'(y)|^2 \, dy \right) dx$$

$$\leq \frac{1}{\beta - 1} \int_0^1 y^\beta |u'(y)|^2 \left(\int_0^y \frac{a(x)}{x^{1+\beta}} \, dx \right) dy.$$

(1.10)

Now, we rewrite

$$\int_0^1 y^\beta |u'(y)|^2 \left(\int_0^y \frac{a(x)}{x^{1+\beta}} \, dx \right) dy = I_\varepsilon + J_\varepsilon + K_\varepsilon,$$

where ε is the positive number given in (1.5),

$$I_\varepsilon = \int_0^\varepsilon y^\beta |u'(y)|^2 \left(\int_0^y \frac{a(x)}{x^{1+\beta}} \, dx \right) dy,$$

$$J_\varepsilon = \int_\varepsilon^1 y^\beta |u'(y)|^2 \left(\int_0^\varepsilon \frac{a(x)}{x^{1+\beta}} \, dx \right) dy,$$

and

$$K_\varepsilon = \int_\varepsilon^1 y^\beta |u'(y)|^2 \left(\int_\varepsilon^y \frac{a(x)}{x^{1+\beta}} \, dx \right) dy.$$

For I_ε, we use (1.5), obtaining

$$\int_0^y \frac{a(x)}{x^{1+\beta}} \, dx \leq \frac{a(y)}{y^\theta} \int_0^y x^{\theta-\beta-1} \, dx \leq \frac{1}{\theta-\beta} a(y) y^{-\beta}.$$

By using this last inequality in I_ε, we deduce that

$$I_\varepsilon \leq \frac{1}{(\theta-\beta)} \int_0^\varepsilon a(x) |u'(x)|^2 \, dx. \tag{1.11}$$

For J_ε, we proceed in a similar way and we get

$$\begin{aligned}
J_\varepsilon &\leq \frac{1}{\theta-\beta} \int_\varepsilon^1 a(y) |u'(y)|^2 \frac{y^\beta}{a(y)} a(\varepsilon) \varepsilon^{-\beta} \, dy \\
&\leq \frac{1}{\theta-\beta} \frac{a(\varepsilon)}{\varepsilon^\beta \inf_{[\varepsilon,1]}(a)} \int_\varepsilon^1 a(x) |u'(x)|^2 \, dx.
\end{aligned} \tag{1.12}$$

For K_ε, we have

$$K_\varepsilon \leq \int_\varepsilon^1 a(y) |u'(y)|^2 \frac{y^\beta}{a(y)} \left(\int_\varepsilon^y \frac{a(x)}{\varepsilon^{1+\beta}} \, dx \right) dy.$$

Therefore,

$$K_\varepsilon \leq \varepsilon^{-1-\beta} \frac{\sup_{[\varepsilon,1]}(a)}{\inf_{[\varepsilon,1]}(a)} \int_\varepsilon^1 a(y) |u'(y)|^2 \, dy. \tag{1.13}$$

By using (1.11)–(1.13) in (1.10), we obtain

$$\int_0^1 \frac{a(x)}{x^2} u^2(x) \, dx \leq C \int_0^1 a(x) |u'(x)|^2 \, dx, \tag{1.14}$$

where the constant C depends on a, ε, θ and β. □

Remark 1.3 1. The previous assertions still hold under different assumptions. In particular, (1.2) holds if we substitute (1.1) and (1.3) with the following hypotheses, respectively,

$$\exists \, \theta \in (0, 1) \text{ s.t. the function } x \mapsto \frac{a(x)}{x^\theta} \text{ is nonincreasing in } (0, 1]$$

and

$\exists\,\theta\in(1,2)$ s.t. the function $x\mapsto\dfrac{a(x)}{x^{\theta}}$ is nondecreasing in $(0,1]$.

In both cases (1.2) holds with the explicit constant $C=\dfrac{4}{(1-\theta)^2}$. Indeed, one can take $\varepsilon=1$ in the proof above, so that $M_{\varepsilon}=N_{\varepsilon}=J_{\varepsilon}=K_{\varepsilon}=0$. Using (1.6) in (1.4) and (1.11) in (1.10) with $\varepsilon=1$, one obtains

$$\int_0^1\frac{a(x)}{x^2}w^2(x)\,dx\le\frac{1}{(\beta-1)(\theta-\beta)}\int_0^1 a(x)|u'(x)|^2\,dx.$$

We remark that this last estimate is optimal for $\beta=\dfrac{\theta+1}{2}$, which gives the desired result.

2. An estimate similar to the previous one is proved in [101] assuming the limit condition

$$\frac{x\,a'(x)}{a(x)}\xrightarrow[x\to0]{}\alpha\quad\text{where }0\le\alpha<2,\tag{1.15}$$

instead of (1.1) or (1.3). We observe that (1.1) and (1.3) in Proposition 1.1 for Hardy–Poincaré type inequalities, and Definition 1.1 or 1.2 for Carleman estimates, are less restrictive than (1.15). Indeed, let us consider $a(x)=x^{\alpha}e^{-\frac{1}{x}}$, with $\alpha\in(1,2)$. Then $xa'(x)/a(x)$ has limit ∞ as x goes to 0^+, whereas for $\alpha\in[1,2)$, (1.3) in Proposition 1.1 holds for any $\theta\in[1,\alpha)$ so that Hardy's inequality holds. Hence the additional hypothesis that $xa'(x)/a(x)$ has a limit $\alpha\in[0,2)$ as x goes to 0^+ in [101] is not optimal for Hardy's inequality.

Moreover, consider the function $a(x)=\sqrt{x}+\epsilon\sqrt{x^3}\sin\dfrac{1}{x}$, for $\epsilon>0$ sufficiently small. This function satisfies Definition 1.1, but $\dfrac{xa'}{a}$ has no limit as $x\to0$, so (1.1) in Proposition 1.1 is less restrictive than the existence of the limit (1.15) also for $\alpha<1$.

We recall here also the classical Hardy inequality that we will use sometimes in the case of Dirichlet boundary conditions:

Proposition 1.2 (Hardy inequality) *There exists $C>0$ such that*

$$\int_0^1\frac{u^2(x)}{x^2}dx\le C\int_0^1(u'(x))^2dx\quad\forall\,u\in H_0^1(0,1).\tag{1.16}$$

See, e.g., [46], for a proof. More in general, we have

Proposition 1.3 (Weighted Hardy inequality) *For all $\alpha\in[0,2)$ and for all $u\in C_c^{\infty}(0,1)$ we have*

$$\frac{(1-\alpha)^2}{4}\int_0^1\frac{u^2(x)}{x^{2-\alpha}}dx\le\int_0^1 x^{\alpha}(u'(x))^2dx.\tag{1.17}$$

Observe that (1.17) ensures that, if $\alpha \in [0, 2) \setminus \{1\}$ and if $u \in H^1_{\text{loc}}(0, 1]$ is such that $x^{\frac{\alpha}{2}} u' \in L^2(0, 1)$, then $\dfrac{u}{x^{\frac{2-\alpha}{2}}} \in L^2(0, 1)$. On the contrary, if $\alpha = 1$, (1.17) does not provide this information anymore. However, a *weaker* Hardy inequality holds in this case (see [46, Chap. 5]):

$$\frac{1}{4} \int_0^1 \frac{u^2(x)}{x(\ln x)} dx \leq \int_0^1 x(u'(x))^2 dx \quad \forall\, u \in C_c^\infty(0, 1). \tag{1.18}$$

In the following we will give some generalizations and improvements of inequality (1.17), also in presence of functions different from the pure power. To do that, we introduce some classes of assumptions, starting from the following one, see [68]:

Hypothesis 1.1 Assume that $a : [0, 1] \to \mathbb{R}_+$ is $C[0, 1] \cap C^1(0, 1]$, $a(0) = 0$, $a > 0$ on $(0, 1]$ and

(i) there exists $K_a \in (0, 2)$ such that

$$\limsup_{x \to 0^+} \frac{xa'(x)}{a(x)} = K_a;$$

(ii) if $K_a \in [1, 2)$, then there exists $m > 0$ and $\delta_0 > 0$ such that for every $x \in (0, \delta_0]$ we have

$$a(x) \geq m \sup_{0 \leq y \leq x} a(y).$$

Clearly, every nondecreasing function near 0 satisfies Hypothesis 1.1.2 with $m = 1$, in particular the power $a(x) = x^{K_a}$, $K_a \geq 1$. Moreover, as an immediate consequence of Hypothesis 1.1.1, we have that $\dfrac{1}{\sqrt{a}} \in L^1(0, 1)$. In particular if $K_a < 1$, then $\dfrac{1}{a} \in L^1(0, 1)$. Notice that Hypothesis 1.1.1 is automatically satisfied when

$$xa'(x) \leq K_a a(x) \text{ for all } x \in (0, 1], \tag{1.19}$$

as assumed in [1, 69], but the former condition is more general than the latter one, see [68].

A set of assumptions similar to Hypothesis 1.1 is given by the following one, see [69]:

Hypothesis 1.2 It is the same as Hypothesis 1.1 with condition (i) replaced by (1.19). $\qquad\qquad\square$

Before giving the generalization of (1.17), we introduce the following spaces:

Definition 1.3 1. If $K_a \in [0, 1)$, set

$$\mathcal{H}^1_a(0, 1) := \{u \in \mathcal{K}^1_a(0, 1) : u(0) = u(1) = 0\},$$

2. if $K_a \in [1, 2)$, set

$$\mathcal{H}_a^1(0, 1) := \{u \in \mathcal{K}_a^1(0, 1) : u(1) = 0\},$$

where $\mathcal{K}_a^1(0, 1)$ is the Hilbert space

$$\mathcal{K}_a^1(0, 1) := \{u \in L^2(0, 1) \cap H^1_{\mathrm{loc}}(0, 1] : a^{\frac{1}{2}} u' \in L^2(0, 1)\}$$

endowed with the scalar product

$$\langle u, v \rangle_{\mathcal{K}_a^1} := \int_0^1 (uv + au'v') dx$$

for all $u, v \in \mathcal{K}_a^1(0, 1)$.

Observe that

1. if $u \in \mathcal{K}_a^1(0, 1)$, the trace of u at $x = 1$ makes sense and this allows to consider Dirichlet condition. On the contrary, the trace at $x = 0$ makes sense only when $K_a \in [0, 1)$. Indeed in this case and using the fact that $\dfrac{1}{a} \in L^1(0, 1)$, we can prove that if $u \in \mathcal{K}_a^1(0, 1)$, then $u \in W^{1,1}(0, 1)$.
2. For $K_a \in [0, 1)$ the space $C_c^\infty(0, 1)$ is dense in $\mathcal{H}_a^1(0, 1)$; on the other hand, if $K_a \in [1, 2)$ the space $\{u \in C^\infty[0, 1] : u(1) = 0\}$ is dense in $\mathcal{H}_a^1(0, 1)$, see [68].

We are now ready to give the promised generalization of (1.17):

Lemma 1.2 (Lemma 18 [69]) *Assume Hypothesis 1.2. Then there exists an optimal constant $\lambda_*(a, K_a)$ such that for every $u \in \mathcal{H}_a^1(0, 1)$, we have*

$$\lambda_*(a, K_a) \int_0^1 \frac{u^2(x)}{x^{2-K_a}} dx \leq \int_0^1 a(x)(u'(x))^2 dx. \tag{1.20}$$

Remark 1.4 In the previous result it is also proved that

$$\lambda_*(a, K_a) \geq a(1) \frac{(1 - \alpha)^2}{4} \geq 0;$$

thus $\lambda_*(a, K_a) > 0$ for every $K_a \in (0, 2) \setminus \{1\}$ but $\lambda_*(a, 1)$ might be zero. Hence if $K_a = 1$, (1.20) is useless. However, as for the power function, a *weaker* Hardy inequality can be proved, in analogy with (1.18):

$$\frac{a(1)}{4} \int_0^1 \frac{u^2(x)}{x \ln x} dx \leq \int_0^1 a(x)(u'(x))^2 dx, \tag{1.21}$$

for every $u \in \mathcal{H}_a^1(0, 1)$, see inequality (21) in [69].

In some cases the following generalizations of (1.17) are useful, as well.

Proposition 1.4 (Theorem 2.1 [115]) *Let $K_a \in [0, 2)$. For all $n > 0$ and $\gamma < 2 - K_a$, there exists a positive constant $C = C(K_a, \gamma, n)$ such that, for all $u \in C_c^\infty(0, 1)$, we have*

$$\frac{(1 - K_a)^2}{4} \int_0^1 \frac{u^2(x)}{x^{2-K_a}} dx + n \int_0^1 \frac{u^2(x)}{x^\gamma} dx \leq \int_0^1 x^{K_a} (u'(x))^2 dx \qquad (1.22)$$
$$+ C \int_0^1 u^2 dx.$$

The precise value of C is given in [115].

Proposition 1.5 (Theorem 2.2 [115]) *Let $K_a \in [0, 2)$. For all $\eta > 0$ there exists a positive constant $C = C(K_a, \eta)$ such that, for all $u \in C_c^\infty(0, 1)$, the following inequality holds:*

$$\frac{(1 - K_a)^2}{4} \int_0^1 \frac{u^2(x)}{x^{2-K_a}} dx + \int_0^1 x^{K_a+\eta} (u'(x))^2 dx \leq C \int_0^1 x^{K_a} (u'(x))^2 dx. \quad (1.23)$$

As for (1.17), one can improve (1.18) and (1.22), obtaining the following two results.

Proposition 1.6 (Theorem 6.3 [115]) *For all $n > 0$ and $\gamma < 1$, there exists $C = C(n, \gamma) > 0$ such that, for all $u \in C_c^\infty(0, 1)$, we have*

$$\frac{1}{4} \int_0^1 \frac{u^2}{x(\ln x)^2} dx + n \int_0^1 \frac{u^2}{x^\gamma} dx \leq \int_0^1 x(u'(x))^2 dx + C \int_0^1 u^2 dx. \qquad (1.24)$$

In the next case, the extension is obtained also when the pure power is replaced by a function a satisfying Hypothesis 1.2.

Proposition 1.7 (Theorem 21 [69]) *Assume Hypothesis 1.2 and let $\lambda < \lambda_*(a, K_a)$. Then, for all $n > 0$ and $\gamma < 2 - K_a$, there exists a constant $C = C(a, K_a, \lambda, n, \gamma) > 0$ such that*

$$\lambda \int_0^1 \frac{u^2}{x^{2-K_a}} dx + n \int_0^1 \frac{u^2}{x^\gamma} dx \leq \int_0^1 a(x)(u'(x))^2 dx + C \int_0^1 u^2 dx \qquad (1.25)$$

for all $u \in \mathcal{H}_a^1(0, 1)$.

Inequalities of this type are the main tool to prove that degenerate/singular problems (like those we have in mind) are well posed (for instance, see [68, Proposition 3 and Theorem 3.4] and the subsequent chapters in these notes), but for that it is essential to generalize Lemma 1.2 and Proposition 1.7 for different weights. More precisely, we have

Lemma 1.3 (Lemma 2.3 [68]) *Assume Hypothesis 1.1, let $K_a \in [0, 2)$ and $\beta \in (0, 2 - K_a)$. Then there exists an optimal constant $\lambda_*(a, \beta)$ such that for every $u \in \mathcal{H}_a^1(0, 1)$, we have*

$$\lambda_*(a, \beta) \int_0^1 \frac{u^2}{x^\beta} dx \leq \int_0^1 a(x)(u'(x))^2 dx. \tag{1.26}$$

Proposition 1.8 (Theorem 2.4 [68]) *Assume Hypothesis 1.1, let $K_a \in [0, 2)$ and $0 < \beta < 2 - K_a$. Then, for all $n > 0$ and $\gamma \in (0, 2 - K_a)$, there exist two constants $C > 0$ and $\mu > 0$ such that*

$$\mu \int_0^1 \frac{u^2}{x^\beta} dx + n \int_0^1 \frac{u^2}{x^\gamma} dx \leq \int_0^1 a(x)(u'(x))^2 dx + C \int_0^1 u^2 dx$$

for all $u \in \mathcal{H}_a^1(0, 1)$.

The last Hardy–Poincaré inequalities that we need to prove well posedness in the purely singular case in presence of N-dimensional ($N \geq 1$) domains are the following ones:

Theorem 1.1 (Theorems 2.1 and 2.2 [118]) *Let Ω be a bounded open subset of $\mathbb{R}^N$, $N \geq 2$, and*

$$\lambda_* := \frac{(N - 2)^2}{4}.$$

Then

- *there exists a constant $C > 0$ such that*

$$C \int_\Omega u^2 dx + \lambda_* \int_\Omega \frac{u^2}{|x|^2} dx \leq \int_\Omega |\nabla u|^2 dx$$

  holds for all $u \in H_0^1(\Omega)$;
- *for any $q \in [1, 2)$, there exists a constant $C = C(q, \Omega) > 0$ such that*

$$C \|u\|_{W^{1,q}(\Omega)}^2 + \lambda_* \int_\Omega \frac{u^2}{|x|^2} dx \leq \int_\Omega |\nabla u|^2 dx$$

  holds for all $u \in H_0^1(\Omega)$.

Similar to Proposition 1.4 is the next inequality, where the degeneracy is assumed to be on the boundary of the domain, see also [57] and [58]:

Proposition 1.9 (Proposition 1.1 [41]) *Let $\Omega \subset \mathbb{R}^N$, $N \geq 1$, be a domain such that $0 \in \partial\Omega$. Then, for any $\lambda \leq \dfrac{N^2}{4}$ and any $\gamma \in [0, 2)$, there exists a constant C depending on γ, λ and Ω, so that every $u \in H_0^1(\Omega)$ satisfies the inequality*

$$\int_\Omega \frac{u^2}{|x|^\gamma} dx + \lambda \int_\Omega \frac{u^2}{|x|^2} dx \le \int_\Omega |\nabla u|^2 dx + C \int_\Omega u^2 dx. \qquad (1.27)$$

The previous inequality can be improved in the following way:

Proposition 1.10 (Proposition 1.2 [41]) *Let $\Omega \subset \mathbb{R}^N$, $N \ge 1$, be a domain such that $0 \in \partial\Omega$. Then, for any $\lambda \le \dfrac{N^2}{4}$ and any $\gamma \in [0, 2)$, there exist two constants C_1, C_2 depending on γ, λ and Ω, so that every $u \in H_0^1(\Omega)$ satisfies the inequality*

$$C_1 \int_\Omega \left(\frac{u^2}{|x|^\gamma} + |x|^{2-\gamma} |\nabla u|^2 \right) dx \le \int_\Omega |\nabla u|^2 dx - \lambda \int_\Omega \frac{u^2}{|x|^2} dx$$
$$+ C_2 \int_\Omega u^2 dx. \qquad (1.28)$$

1.2 Some Hardy–Poincaré Inequalities with Interior Degeneracy

As announced, we are also interested in degeneracy or singularity occurring in the interior of the spatial domain. For this reason, a suitable functional framework is necessary. We recall that degenerate problems will be treated only in the 1-dimensional case, so in this case we will assume $\Omega = (0, 1)$ and $x_0 \in (0, 1)$.

Let us start considering the variant of Proposition 1.1 which is needed when a degenerates at an interior point x_0:

Proposition 1.11 (Proposition 2.6 [77]) *Assume that $a \in C([0, 1])$, $a > 0$ in $[0, 1] \setminus \{x_0\}$, $a(x_0) = 0$ and there exists $q > 1$ such that the function*

$$x \mapsto \frac{a(x)}{|x - x_0|^q} \text{ is nonincreasing on the left of } x = x_0 \qquad (1.29)$$
$$\text{and nondecreasing on the right of } x = x_0.$$

Then there exists a constant $C > 0$ such that for any function u, locally absolutely continuous on $[0, x_0) \cup (x_0, 1]$ and satisfying

$$u(0) = u(1) = 0 \text{ and } \int_0^1 a(x)|u'(x)|^2 \, dx < +\infty,$$

the following inequality holds:

$$\int_0^1 \frac{a(x)}{(x - x_0)^2} u^2(x) \, dx \le C \int_0^1 a(x)|u'(x)|^2 \, dx. \qquad (1.30)$$

Actually, such a proposition was stated in [77] requiring $1 < q < 2$. However, as it is clear from the proof below, the result is true without the upper bound on q, that in [77] was used for other estimates.

Moreover, we also remark that this proposition is the interior counterpart of Proposition 1.1, but only in the case $q > 1$. Indeed, the analogous of the case $q \in (0, 1)$ therein cannot hold here, since we are missing the information "$u(x_0) = 0$", which would correspond to the assumption "$u(0) = 0$" made in Proposition 1.1.

Although, as already said, we shall treat only Dirichlet problems, we mention that analogous Hardy–Poincaré inequalities in the Neumann case are available, for instance see [70, Proposition 2.2], [71, Proposition 2.2] or [71, Corollary 2.1].

Proof (*Proof of Proposition* 1.11) Fix any $\beta \in (1, q)$ and $\varepsilon > 0$ small. Since $u(1) = 0$, applying Hölder's inequality and Fubini's Theorem, we have

$$
\begin{aligned}
&\int_{x_0+\varepsilon}^{1} \frac{a(x)}{(x-x_0)^2} u^2(x)\, dx \\
&= \int_{x_0+\varepsilon}^{1} \frac{a(x)}{(x-x_0)^2} \left(\int_x^1 ((y-x_0)^{\beta/2} u'(y))(y-x_0)^{-\beta/2}\, dy \right)^2 dx \\
&\leq \int_{x_0+\varepsilon}^{1} \frac{a(x)}{(x-x_0)^2} \left(\int_x^1 (y-x_0)^{\beta} |u'(y)|^2\, dy \int_x^1 (y-x_0)^{-\beta}\, dy \right) dx \\
&\leq \frac{1}{\beta-1} \int_{x_0+\varepsilon}^{1} \frac{a(x)}{(x-x_0)^{1+\beta}} \left(\int_x^1 (y-x_0)^{\beta} |u'(y)|^2\, dy \right) dx \\
&= \frac{1}{\beta-1} \int_{x_0+\varepsilon}^{1} (y-x_0)^{\beta} |u'(y)|^2 \left(\int_{x_0+\varepsilon}^{y} \frac{a(x)}{(x-x_0)^{1+\beta}}\, dx \right) dy \\
&= \frac{1}{\beta-1} \int_{x_0+\varepsilon}^{1} (y-x_0)^{\beta} |u'(y)|^2 \left(\int_{x_0+\varepsilon}^{y} \frac{a(x)}{(x-x_0)^{q}} (x-x_0)^{q-1-\beta}\, dx \right) dy.
\end{aligned}
$$

Thanks to our hypothesis, we find

$$
\frac{a(x)}{(x-x_0)^q} \leq \frac{a(y)}{(y-x_0)^q}, \quad \forall\, x,\, y \in [x_0+\varepsilon, 1],\ x < y.
$$

Thus

$$
\begin{aligned}
&\int_{x_0+\varepsilon}^{1} \frac{a(x)}{(x-x_0)^2} u^2(x)\, dx \\
&\leq \frac{1}{\beta-1} \int_{x_0+\varepsilon}^{1} \frac{a(y)}{(y-x_0)^q} (y-x_0)^{\beta} |u'(y)|^2 \left(\int_{x_0+\varepsilon}^{y} (x-x_0)^{q-1-\beta}\, dx \right) dy \\
&= \frac{1}{(\beta-1)(q-\beta)} \int_{x_0+\varepsilon}^{1} \frac{a(y)}{(y-x_0)^q} (y-x_0)^{q-\beta} (y-x_0)^{\beta} |u'(y)|^2\, dy \\
&= \frac{1}{(\beta-1)(q-\beta)} \int_{x_0+\varepsilon}^{1} p(y) |u'(y)|^2\, dy.
\end{aligned}
\tag{1.31}
$$

Now, proceeding as before, and using the fact that $u(0) = 0$, one has

$$\int_0^{x_0-\varepsilon} \frac{a(x)}{(x_0-x)^2} u^2(x)\,dx$$

$$\leq \int_0^{x_0-\varepsilon} \frac{a(x)}{(x_0-x)^2} \left(\int_0^x (x_0-y)^\beta |u'(y)|^2\,dy \int_0^x (x_0-y)^{-\beta}\,dy \right) dx$$

$$\leq \frac{1}{\beta-1} \int_0^{x_0-\varepsilon} \frac{a(x)}{(x_0-x)^{1+\beta}} \left(\int_0^x (x_0-y)^\beta |u'(y)|^2\,dy \right) dx$$

$$= \frac{1}{\beta-1} \int_0^{x_0-\varepsilon} (x_0-y)^\beta |u'(y)|^2 \left(\int_y^{x_0-\varepsilon} \frac{a(x)}{(x_0-x)^q} (x_0-x)^{q-1-\beta}\,dx \right) dy.$$

Again by assumption

$$\frac{a(x)}{(x_0-x)^q} \leq \frac{a(y)}{(x_0-y)^q}, \quad \forall\, x,\, y \in [0, x_0-\varepsilon],\ y < x.$$

Hence,

$$\int_0^{x_0-\varepsilon} \frac{a(x)}{(x_0-x)^2} u^2(x)\,dx$$

$$\leq \frac{1}{\beta-1} \int_0^{x_0-\varepsilon} \frac{a(y)}{(x_0-y)^q} (x_0-y)^\beta |u'(y)|^2 \left(\int_y^{x_0-\varepsilon} (x_0-x)^{q-1-\beta}\,dx \right) dy$$

$$= \frac{1}{(\beta-1)(q-\beta)} \int_0^{x_0-\varepsilon} a(y)|u'(y)|^2\,dy.$$

$$(1.32)$$

Passing to the limit as $\varepsilon \to 0$ and combining (1.31) and (1.32), the conclusion follows.
□

In order to establish the functional setting we shall use, we start introducing the reference Hilbert spaces depending on two general functions a and b degenerating at an interior point $x_0 \in (0, 1)$: given a and b satisfying **(WD)** or **(SD)**, we introduce the weighted Hilbert spaces

$$\mathcal{H}^1_{a,x_0}(0, 1) := \left\{ u \in W_0^{1,1}(0, 1) \,:\, \sqrt{a}\,u' \in L^2(0, 1) \right\}$$

and

$$\mathcal{H}^1_{a,b,x_0}(0, 1) := \left\{ u \in \mathcal{H}^1_{a,x_0}(0, 1) \,:\, \frac{u}{\sqrt{b}} \in L^2(0, 1) \right\},$$

endowed with the inner products

$$\langle u, v \rangle_{\mathcal{H}^1_{a,x_0}(0,1)} := \int_0^1 au'v'\,dx + \int_0^1 uv\,dx,$$

and

$$\langle u, v \rangle_{\mathcal{H}^1_{a,b,x_0}(0,1)} = \int_0^1 au'v'dx + \int_0^1 uv\, dx + \int_0^1 \frac{uv}{b}dx,$$

respectively.

Notice that, if $u \in \mathcal{H}^1_{a,x_0}(0,1)$, then $au' \in L^2(0,1)$, since

$$|au'| \le (\max_{[0,1]} \sqrt{a})\sqrt{a}|u'|.$$

Lemma 1.4 (Lemma 2.9 [79]) *If $K_a + K_b \le 2$ and $K_b < 1$, then there exists a constant $C > 0$ such that*

$$\int_0^1 \frac{u^2}{b}dx \le C \int_0^1 a(u')^2 dx \tag{1.33}$$

for every $u \in \mathcal{H}^1_{a,x_0}(0,1)$.

Proof We set $p(x) := \dfrac{(x-x_0)^2}{b}$, so that p satisfies (1.29) with $q = 2 - K_b > 1$ by Lemma 1.1. Thus, taken $u \in \mathcal{H}^1_{a,x_0}(0,1)$, by Proposition 1.11, we get

$$\int_0^1 \frac{u^2}{b}dx = \int_0^1 \frac{a(x)}{(x-x_0)^2}u^2 dx \le C \int_0^1 a(x)|u'(x)|^2 dx.$$

Now, by Lemma 1.1,

$$p(x) = (x-x_0)^{2-K_a-K_b}a(x)\frac{(x-x_0)^{K_a}}{a(x)}\frac{(x-x_0)^{K_b}}{b(x)} \le ca(x)$$

for some $c > 0$, and the claim follows. $\qquad\square$

Remark 1.5 A similar proof shows that, when $K_a + 2K_b \le 2$ and $K_b < 1/2$, then

$$\int_0^1 \frac{u^2}{b^2}dx \le C \int_0^1 a(u')^2 dx$$

for every $u \in \mathcal{H}^1_{a,x_0}(0,1)$.

Lemma 1.4 implies that $\mathcal{H}^1_{a,x_0}(0,1) = \mathcal{H}^1_{a,b,x_0}(0,1)$ when $K_a + K_b \le 2$ and $K_b < 1$. However, inequality (1.33) holds in other cases, see Proposition 1.12 below. In order to prove such a proposition, we need a preliminary result.

Lemma 1.5 *If $K_b \ge 1$, then $u(x_0) = 0$ for every $u \in \mathcal{H}^1_{a,b,x_0}(0,1)$.*

Proof Since $u \in W^{1,1}_0(0,1)$, there exists $\lim_{x \to x_0} u(x) = L \in \mathbb{R}$. If $L \ne 0$, then $|u(x)| \ge \dfrac{L}{2}$ in a neighborhood of x_0, that is

$$\frac{|u(x)|^2}{b} \geq \frac{L^2}{4b} \notin L^1(0, 1)$$

by Lemma 1.1, and thus $L = 0$. $\qquad\qquad\qquad\qquad\qquad\qquad\qquad\qquad\qquad\qquad\quad\square$

We also need the following result, whose proof, with the aid of Lemma 1.5, is a simple adaptation of the one given in [84, Lemma 3.2].

Lemma 1.6 *If $K_b \geq 1$, then*

$$H_c^1(0, 1) := \left\{ u \in H_0^1(0, 1) \text{ such that supp } u \subset (0, 1) \setminus \{x_0\} \right\}$$

is dense in $\mathcal{H}_{a,b,x_0}^1(0, 1)$.

In the spirit of [46, Lemma 5.3.1], now we are ready for the following "classical" weighted Hardy inequality in the space $\mathcal{H}_{a,b,x_0}^1(0, 1)$ for $a(x) = |x - x_0|^\alpha$ and $b(x) = |x - x_0|^{2-\alpha}$, proved in [79, Lemma 2.13]. However, note that the inequality below is different from the classical one, since we admit a singularity inside the interval:

Lemma 1.7 (Weighted Hardy inequality with interior degeneracy) *For every $\alpha \in \mathbb{R}$ the inequality*

$$\frac{(1-\alpha)^2}{4} \int_0^1 \frac{u^2}{|x - x_0|^{2-\alpha}} dx \leq \int_0^1 |x - x_0|^\alpha (u')^2 dx$$

holds true for every $u \in \mathcal{H}_{|x-x_0|^\alpha, |x-x_0|^{2-\alpha}, x_0}^1(0, 1)$.

Proof The case $\alpha = 1$ is trivial. So, take $\beta = (1 - \alpha)/2 \neq 0$ and $\varepsilon \in (0, 1 - x_0)$. First case: $\beta < 0$ ($\alpha > 1$). In this case we have

$$\int_{x_0+\varepsilon}^1 (x - x_0)^\alpha (u')^2 dx$$

$$= \int_{x_0+\varepsilon}^1 (x - x_0)^\alpha \left((x - x_0)^\beta \left((x - x_0)^{-\beta} u \right)' + \beta(x - x_0)^{-1} u \right)^2 dx$$

$$\geq \beta^2 \int_{x_0+\varepsilon}^1 (x - x_0)^{\alpha-2} u^2 dx + 2\beta \int_{x_0+\varepsilon}^1 (x - x_0)^{\alpha+\beta-1} u \left((x - x_0)^{-\beta} u \right)' dx$$

$$= \beta^2 \int_{x_0+\varepsilon}^1 (x - x_0)^{\alpha-2} u^2 dx + \beta \left((x - x_0)^{-\beta} u \right)^2 \Big|_{x_0+\varepsilon}^1$$

(since $\alpha + \beta - 1 = -\beta$)

$$\geq \beta^2 \int_{x_0+\varepsilon}^1 (x - x_0)^{\alpha-2} u^2 dx.$$

Letting $\varepsilon \to 0^+$, we get that

$$\int_{x_0}^1 (x - x_0)^\alpha (u')^2 dx \geq \beta^2 \int_{x_0}^1 (x - x_0)^{\alpha-2} u^2 dx. \tag{1.34}$$

Second case: $\beta > 0$. In this situation we have $2 - \alpha > 1$. Thus, in view of Lemma 1.6 with $K_b = 2 - \alpha$, we will prove (1.34) first if $u \in H_c^1(0, 1)$ and then, by density, if $u \in H_{|x-x_0|^\alpha, |x-x_0|^{2-\alpha}, x_0}^1(0, 1)$. Thus, take $u \in H_c^1(0, 1)$; proceeding as above, we get

$$\int_{x_0+\varepsilon}^1 (x - x_0)^\alpha (u')^2 dx$$

$$\geq \beta^2 \int_{x_0+\varepsilon}^1 (x - x_0)^{\alpha-2} u^2 dx + \beta \big((x - x_0)^{-\alpha} u\big)^2 \Big|_{x_0+\varepsilon}^1$$

$$\geq \beta^2 \int_{x_0+\varepsilon}^1 (x - x_0)^{\alpha-2} u^2 dx,$$

since $u(x_0 + \varepsilon) = 0$ for ε small enough.

Passing to the limit as $\varepsilon \to 0^+$, and using Lemma 1.6, we get that (1.34) holds true for every $u \in H_{|x-x_0|^\alpha, |x-x_0|^{2-\alpha}, x_0}^1(0, 1)$.

Operating in a symmetric way on the left of x_0, we get the conclusion. $\qquad\square$

Remark 1.6 We remark that the previous inequality is valid *for every* $\alpha \in \mathbb{R}$. This means, for example, that, when $\alpha < 0$, we have an inequality with two singular weights.

As a corollary of the previous result, we get the following improvement of Lemma 1.4.

Proposition 1.12 (Proposition 2.14 [79]) *If a and b are* **(WD)** *and/or* **(SD)** *with $K_a + K_b \leq 2$, then* (1.33) *holds for every* $u \in \mathcal{H}_{a,b,x_0}^1(0, 1)$.

Remark 1.7 It is well known that when $K_a = K_b = 1$, an inequality of the form (1.33) doesn't hold (see [106]). Being such an inequality fundamental for the observability inequality (1.45), it is no surprise if with our techniques we cannot handle this case in Chap. 5. We underline also that Lemma 1.4 and Proposition 1.12 are proved in [79] under weaker assumptions on the functions a and b.

The fundamental space in which we will work is clearly the one where the Hardy–Poincaré–type inequality (1.33) holds: in view of Proposition 1.12, it is clear that such a space is

$$\mathcal{H}_0 := \mathcal{H}_{a,b,x_0}^1(0, 1). \tag{1.35}$$

Remark 1.8 Under the assumptions of Proposition 1.12, the standard norm $\|\cdot\|_{\mathcal{H}_0}^2$ is equivalent to

$$\|u\|_\circ^2 := \int_0^1 a(u')^2 dx$$

for all $u \in \mathcal{H}_0$. Indeed, for all $u \in \mathcal{H}_0$, we have

$$\int_0^1 u^2 dx = \int_0^1 b \frac{u^2}{b} dx \le c \int_0^1 a(u')^2 dx,$$

and this is enough to conclude.

Moreover, when $\lambda < 0$, an equivalent norm is obviously given by

$$\|u\|_{\sim}^2 := \int_0^1 a(u')^2 dx - \lambda \int_0^1 \frac{u^2}{b} dx.$$

The previous Hardy–Poincaré inequalities are crucial for the problem in divergence form. For the degenerate/singular problem in non divergence form, we need different weighted Hardy–Poincaré inequalities that will be very important in Sect. 5.2. In order to deal with these inequalities we consider different classes of weighted Hilbert spaces, which are suitable to study the cases when a and b are **(SD)** or **(WD)**. Thus, we introduce

$$\mathcal{H}_{\frac{1}{a}, x_0}^1 (0, 1) := L_{\frac{1}{a}}^2 (0, 1) \cap H_0^1 (0, 1)$$

and

$$\mathcal{K}_{a,b,x_0}(0, 1) := \left\{ u \in \mathcal{H}_{\frac{1}{a}, x_0}^1 (0, 1) : \frac{u}{\sqrt{ab}} \in L^2(0, 1) \right\}$$

with the inner products

$$\langle u, v \rangle_{\mathcal{H}_{\frac{1}{a}, x_0}^1} = \int_0^1 \frac{uv}{a} dx + \int_0^1 u'v' dx,$$

and

$$\langle u, v \rangle_{\mathcal{K}_{a,b,x_0}} = \int_0^1 \frac{uv}{a} dx + \int_0^1 u'v' dx + \int_0^1 \frac{uv}{ab} dx,$$

respectively. Here, as usual, we have denoted by $L_{\frac{1}{a}}^2 (0, 1)$ the L^2 space with respect to the weight $\frac{1}{a}$ in $(0, 1)$.

We start making the following assumption:

Hypothesis 1.3 1. a and b are **(WD)** with $K_a + K_b < 1$, or
2. a and b are **(WD)** with $1 \le K_a + K_b \le 2$ and

$$\exists\, c_1, c_2 > 0 \text{ such that } |x - x_0|^{K_a} \ge c_1 a \text{ and } |x - x_0|^{K_b} \ge c_2 b \qquad (1.36)$$

for all $x \in [0, 1]$, or
3. a is **(WD)** and b is **(SD)** or a is **(SD)** and b is **(WD)** with $K_a + K_b \le 2$ and (1.36), or
4. a and b are **(SD)** with $K_a = K_b = 1$. $\square$

Observe that (1.36) is quite natural. Indeed, if we consider the prototype functions $a(x) = |x - x_0|^{K_a}$ and $b(x) = |x - x_0|^{K_b}$, with $K_a + K_b \geq 1$, (1.36) is clearly satisfied with $c_1 = c_2 = 1$.

Using these weighted spaces we can prove the next Hardy–Poincaré inequalities. Indeed, thanks to Lemma 1.1 and Proposition 1.11, one can prove the following estimate:

Lemma 1.8 (Lemma 2.4 [71]) *Assume that Hypothesis 1.3.1 holds. Then there exists a constant $C > 0$ such that*

$$\int_0^1 \frac{u^2}{ab} dx \leq C \int_0^1 (u')^2 dx \qquad (1.37)$$

for every $u \in \mathcal{H}_{\frac{1}{a}, x_0}(0, 1)$.

In the other cases, it holds:

Lemma 1.9 (Lemma 2.5 [71]) *Assume that one among Hypothesis 1.3.2, 1.3.3 or 1.3.4 is satisfied. Then there exists a constant $C > 0$ such that (1.37) holds for every $u \in \mathcal{K}_{a,b,x_0}(0, 1)$.*

Observe that the previous estimates give Hardy–Poincaré inequalities in all situations; however, Lemma 1.9 allows us to consider only the case when a and b are both **(SD)** with $K_a = K_b = 1$. Moreover, thanks to Lemma 1.8, if $K_a + K_b < 1$, the spaces $\mathcal{H}_{\frac{1}{a}, x_0}(0, 1)$ and $\mathcal{K}_{a,b,x_0}(0, 1)$ coincide and the two norms are equivalent. Thus, define

$$\mathcal{K} := \begin{cases} \mathcal{H}_{\frac{1}{a}, x_0}(0, 1), & \text{if Hypothesis 1.3.1 is satisfied,} \\ \mathcal{K}_{a,b,x_0}(0, 1), & \text{if Hypothesis 1.3.2, 1.3.3 or 1.3.4 is in force,} \end{cases} \qquad (1.38)$$

so that the Hardy–Poincaré inequality (1.37) holds in $\mathcal{K}$. We underline also that, if the assumptions of Lemma 1.8 or 1.9 are satisfied, then the standard norm $\| \cdot \|_{\mathcal{K}}$ is equivalent to

$$\|u\|_1^2 := \int_0^1 (u'(x))^2 dx$$

for all $u \in \mathcal{K}$. Indeed, if (1.37) holds, for all $u \in \mathcal{K}$, we have

$$\int_0^1 \frac{u^2}{a} dx = \int_0^1 b \frac{u^2}{ab} dx \leq c \int_0^1 (u')^2 dx,$$

for a positive constant c, and this is enough to conclude.

Remark 1.9 The introduction of the space $\mathcal{K}$ made in [77] is new with respect to all the previous approaches: including the integrability of u^2/b in the definition of $\mathcal{H}$ has the advantage of obtaining immediately some useful functional properties, that in general could be hard to show in the usual Sobolev spaces. Indeed, solutions were already found in suitable function spaces for the "critical" and "supercritical" cases (when λ equals or exceeds the best constant in the classical Hardy–Poincaré inequality) in [116, 118] for purely singular problems. However, as already done in the purely degenerate case [1, 19, 20, 29, 30, 36, 68–70, 77, 78, 84], a weighted Sobolev space must be used. For this reason, we believe that it is natural to unify these approaches in the singular/degenerate, as we do.

For analogous results in the Neumann case we refer to [71, 82].

1.3 Null Controllability and Observability Inequality

It is well known that the notion of null controllability is equivalent to the one of observability for the associated adjoint problem. As far as we know, such an equivalence has been recognized since the late '60s of the XX century (for instance, see [60, 104]) and has been formally stated in an infinite dimensional setting in [49], essentially adapting theorems in functional analysis which relate the range of an operator and the null space of its adjoint operator. We shall present a proof of this equivalence hereafter, using a classical dual approach, consisting in considering adjoint problems.

To this aim, we take a domain $\Omega \subset \mathbb{R}^N$, $N \geq 1$, and we consider the Hilbert space $X = L^2(\Omega, \mu)$ for some underlying measure μ. This generality lets us consider problems which can be degenerate, singular, with weights, governed by operators in divergence or non divergence form, etc. in a unique setting. Thus, take a linear self-adjoint operator

$$\mathcal{A} : D(\mathcal{A}) \subset X \to X,$$

and consider the problem

$$\begin{cases} u_t - \mathcal{A}u = f\chi_\omega, & (t, x) \in Q := (0, T) \times \Omega, \\ Bu_{|\partial\Omega} = 0, & t \in (0, T), \\ u(0, x) = u_0(x) \in X, & x \in \Omega, \end{cases} \tag{1.39}$$

where B is a boundary operator (typically, Dirichlet or Neumann), χ_ω is the characteristic function of the subdomain $\omega \subset \Omega$ and $f \in L^2(0, T; X)$. Of course, our prototype will be operators of the form

$$\mathcal{A}u = \Delta u, \ \mathcal{A}u = (a(x)u_x)_x, \ \mathcal{A}u = a(x)u_{xx},$$

$$\mathcal{A}u = \Delta u + \lambda\frac{u}{b}, \ \mathcal{A}u = (au_x)_x + \frac{\lambda}{b}u, \ \mathcal{A}u = au_{xx} + \frac{\lambda}{b}u,$$

where a and b can be strictly positive or can degenerate somewhere.

In an abstract framework, problem (1.39) can be formalized in the following way (see the "Hilbert Uniqueness Method" (HUM) in [96]): let us denote by V an intermediate Sobolev space, $D(\mathcal{A}) \subset V \subset X$, with continuous and dense embeddings. In the examples above, $X = L^2(\Omega, \mu)$, V is a Sobolev space of order 1 and $D(\mathcal{A})$ a Sobolev space of order 2. Let us suppose that in the previous setting, for every $u_0 \in X$ and every $f \in L^2(0, T; X)$ problem (1.39) has a unique solution $u \in L^2(0, T; V) \cap H^1(0, T; V^*) \cap C([0, T]; X)$, where V^* is the dual of V. Here, we have in mind the following definition.

Definition 1.4 We say that u is a (weak) solution to problem (1.39) if

$$u \in L^2(0, T; V) \cap H^1(0, T; V^*)$$

and

$$\langle u(T), \phi(T) \rangle_X - \langle u_0, \phi(0) \rangle_X - \int_0^T \langle v_t, u \rangle_{V^*, V} dt - \int_0^T \langle \mathcal{A}u, \phi \rangle_{V^*, V} dt$$
$$= \int_0^T \langle f \chi_\omega, \phi \rangle_X dt$$

for every $\phi \in L^2(0, T; V) \cap H^1(0, T; V^*)$.

Note that, by [98, Chap. 3.4.4], any solution of (1.39) belongs to $L^2(0, T; V^*)$ $\subset C([0, T]; X)$,[1] and so the functional setting is meaningful.

Moreover, a classical parabolic estimate will be assumed (as it holds in the standard cases, see [45, Chap. XVIII]):

$$\|u\|_{C([0,T];X)}^2 + \|u\|_{L^2(0,T;V)}^2 + \|u\|_{H^1(0,T;V^*)}^2 \leq C \left(\|u_0\|_X^2 + \|f\|_{L^2(0,T;X)}^2 \right) \quad (1.40)$$

for some universal constant $C > 0$.

Remark 1.10 If u is a solution to problem (1.39), we have that $u(t, \cdot) \in V$ for a.e $t \in [0, T]$. But a stronger regularity result holds. Indeed, the following result holds true:

Theorem 1.2 ([109], Corollary IV.6.4) *Let $a : V \times V \to \mathbb{R}$ defined as $a(u, v) :=$ $\langle \mathcal{A}u, v \rangle_{V^*, V}$ be continuous and coercive, that is*

$$|a(u, v)| \leq c_1 \|u\|_V \|v\|_V \quad and \quad a(u, u) \geq c_2 \|u\|_V^2$$

for all $u, v \in V$ and for some $c_1, c_2 > 0$. Then $D(\mathcal{A})$ is dense in X and for every $u_0 \in X$ there is a unique solution $u \in C([0, T]; X) \cap C^\infty((0, T]; X)$ of (1.39) when $f = 0$. Moreover, $u(t) \in D(\mathcal{A}^p)$ for each $t > 0$ and for every $p \geq 1$.

---

[1] Indeed, by [98, Theorem 1.3.1, Remarks 1.3.1 and 1.2.4] we have that $u' \in C([0, T]; [V, X]_{\frac{3}{4}}) \subset C([0, T]; [V, X]_1) = C([0, T]; X)$, where $[V, X]_\theta$ denotes the θ-th intermediate space between V and X.

Actually, some more regularity in terms of Bochner spaces can be obtained: since $u(t, \cdot) \in V$ for a.e $t \in [0, T]$, let t_0 be one of such points and consider the problem

$$
\begin{cases}
z_t - \mathcal{A}z = f\chi_\omega, & t > t_0, \\
Bz_{|\partial\Omega} = 0, & t \in (t_0, T), \\
z(t_0) = u(t_0) \in V.
\end{cases}
\tag{1.41}
$$

Assuming that $D(\mathcal{A})$ is dense in X, by [99, Chap. 4, Theorem 3.2],[2] we have that $z \in L^2(t_0, T; D(\mathcal{A}))$. On the other hand, by uniqueness, $z = u$ for any $t \geq t_0$. Moreover, problem (1.41) can be formulated for almost any $t_0 \in [0, T]$, and so $u \in L^2(\epsilon, T; D(\mathcal{A}))$ for every $\epsilon > 0$.

We start recalling some basic definitions.

Definition 1.5 Problem (1.39) is *(globally) null controllable*, (GNC) for shortness, if for all $T > 0$ and for all initial datum $u_0 \in X$ there exists a control $f \in L^2(0, T; X)$ such that the solution of problem (1.39) satisfies

$$
u(T, x) = 0 \quad \text{for all } x \in \Omega
\tag{1.42}
$$

and there exists a positive constant C such that

$$
\|f\|_{L^2(0,T;X)} \leq C\|u_0\|_X.
\tag{1.43}
$$

Finally, we consider the homogeneous adjoint problem to (1.39):

$$
\begin{cases}
v_t + \mathcal{A}^*v = 0, & (t, x) \in Q, \\
Bv_{|\partial\Omega} = 0, & t \in (0, T), \\
v(T, x) = v_T(x) \in X, & x \in \Omega.
\end{cases}
\tag{1.44}
$$

Definition 1.6 We say that problem (1.44) is *observable* if there exists a constant $C = C(\mathcal{A}, T; \omega) > 0$ such that for all $v_T \in X$ the solution v of (1.44) satisfies

$$
\int_\Omega v^2(0, x)d\mu \leq C \int_0^T \int_\omega v^2(t, x)d\mu dt.
\tag{1.45}
$$

For the sake of simplicity, it is commonly said that problem (1.39) is null controllable and that problem (1.44) satisfies the observability inequality (1.45).

We are now ready for the equivalence between the two notions.

---

[2]In the cited theorem $u_0 \in [D(\mathcal{A}), X]_{\frac{1}{2}}$, but since $D(\mathcal{A})$ is dense in X we have that $[D(\mathcal{A}), X]_{\frac{1}{2}} = V$, see [98, Chap. 1, Eq. (2.42)].

Theorem 1.3 *Assume* (1.40). *Problem* (1.39) *is null controllable if and only if problem* (1.44) *satisfies the observability inequality* (1.45).

Proof First, assume that (1.39) is null controllable. Take $u_0(x) := v(0, x)$ as initial value; then use v as test function in (1.39) and u as test function for (1.44). Summing up we find

$$\int_\Omega u(T, x)v(T, x)d\mu - \int_\Omega u(0, x)v(0, x)d\mu = \int_Q f v \chi_\omega d\mu dt. \qquad (1.46)$$

By assumption, $u(T, x) = 0$ in Ω and $u(0, x) = u_0(x) = v(0, x)$; hence, taking $\epsilon > 0$, by using the Schwarz inequality, (1.46) becomes

$$\int_\Omega v^2(0, x)d\mu = -\int_Q f v \chi_\omega d\mu dt = -\int_0^T \int_\omega f v d\mu dt$$
$$\leq \frac{\epsilon}{2} \int_0^T \int_\omega f^2 d\mu dt + \frac{1}{2\epsilon} \int_0^T \int_\omega v^2 d\mu dt$$

(by (1.43))

$$\leq \frac{\epsilon C}{2} \int_\Omega u_0^2 d\mu + \frac{1}{2\epsilon} \int_0^T \int_\omega v^2 d\mu dt$$
$$= \frac{\epsilon C}{2} \int_\Omega v^2(0, x)d\mu + \frac{1}{2\epsilon} \int_0^T \int_\omega v^2 d\mu dt.$$

Hence

$$\left(1 - \frac{\epsilon C}{2}\right) \int_\Omega v^2(0, x)d\mu \leq \frac{1}{2\epsilon} \int_0^T \int_\omega v^2 d\mu dt.$$

By choosing $\epsilon < 2/C$, we get that (1.45) holds true.

Now, assume that (1.44) satisfies the observability inequality (1.45). Fixed $\epsilon > 0$ and $u_0 \in X$, define $J_\epsilon : X \to \mathbb{R}$ by

$$J_\epsilon(v_T) = \frac{1}{2} \int_0^T \int_\omega v^2 d\mu dt + \epsilon \|v_T\|_X + \int_\Omega v(0, x)u_0(x)d\mu,$$

where v is the solution to (1.44) with final datum v_T. Since problem (1.44) is linear, we have that J_ϵ is strictly convex (see [56] for a detailed study of $J_\epsilon{}^3$). Moreover, it is continuous and coercive. Thus, there exists a unique minimizer v_T^ϵ of J_ϵ. Let v_ϵ be the solution of (1.44) with $v_\epsilon(T, x) = v_T^\epsilon(x)$ for every $x \in \Omega$, and define $f_\epsilon := v_\epsilon \chi_\omega$. Now, let u_ϵ be the solution of (1.39) with right-hand side f_ϵ, that is

---

[3] With the approach therein, a clear application of the unique continuation principle is presented.

$$\begin{cases} (u_\epsilon)_t - \mathcal{A}u_\epsilon = f_\epsilon = v_\epsilon \chi_\omega, & (t, x) \in Q, \\ Bu_{\epsilon|\partial\Omega} = 0, & t \in (0, T), \\ u_\epsilon(0, x) = u_0(x) \in X, & x \in \Omega. \end{cases} \quad (1.47)$$

First, we note that if for some ϵ_0 we have $v_T^{\epsilon_0} = 0$, then $v_{\epsilon_0} = 0$ and so $J_\epsilon(v_T^{\epsilon_0}) = 0$. As a consequence, for any $v_T \in X$ we have

$$0 \le \lim_{t\to 0^+} \frac{J_\epsilon(tv_T)}{t} = \epsilon\|v_T\|_X + \int_\Omega v(0, x)u_0(x)d\mu,$$

so that

$$\epsilon\|v_T\|_X \ge -\int_\Omega v(0, x)u_0(x)d\mu. \quad (1.48)$$

This implies that we can assume that there exists $\epsilon_0 > 0$ such that

$$\min J_\epsilon < 0 \qquad \text{for all } \epsilon < \epsilon_0.$$

Indeed, assume by contradiction that there exists a sequence $\epsilon_n \downarrow 0$ such that $\min J_{\epsilon_n} = 0$ for every $n \in \mathbb{N}$. Then, by (1.48) we would get

$$-\int_\Omega v(0, x)u_0(x)d\mu \le \epsilon_n\|v_T\|_X \qquad \text{for all } n \in \mathbb{N} \text{ and all } v_T \in X.$$

Then we would get

$$-\int_\Omega v(0, x)u_0(x)d\mu \le 0 \qquad \text{for all } v_T \in X$$

and so

$$\int_\Omega v(0, x)u_0(x)d\mu = 0 \qquad \text{for all } v_T \in X. \quad (1.49)$$

Now, $u_{\epsilon_n} = u$ solves

$$\begin{cases} u_t - \mathcal{A}u = 0, & (t, x) \in Q, \\ Bu_{|\partial\Omega} = 0, & t \in (0, T), \\ u(0, x) = u_0(x) \in X, & x \in \Omega. \end{cases}$$

Using v, solution of (1.44), as test function in the previous problem and u as test function in (1.44), by (1.49) we get

$$\int_\Omega u(T, x)v_T(x)\,d\mu = 0 \qquad \text{for all } v_T \in X,$$

and so $u(T, x) = 0$ in Ω.

Thus, we can suppose that $v_T^\epsilon \neq 0$, so that J_ϵ is differentiable at v_T^ϵ. Since v_T^ϵ is the minimum point of J_ϵ, for all $v_T \in X$ we have

$$
\begin{aligned}
0 &= \left[\frac{d}{dt} J_\epsilon(v_T^\epsilon + t v_T) \right]_{t=0} \\
&= \int_0^T \int_\omega v_\epsilon v \, d\mu \, dt + \epsilon \frac{\int_\Omega v_T^\epsilon v_T \, d\mu}{\|v_T^\epsilon\|_X} + \int_\Omega v(0, x) u_0(x) \, d\mu,
\end{aligned}
\tag{1.50}
$$

where v solves (1.44) with $v(T, x) = v_T(x)$. In particular, putting $v_T = v_T^\epsilon$ in (1.50), one has

$$
\int_0^T \int_\omega v_\epsilon^2 \, d\mu \, dt = -\epsilon \|v_T^\epsilon\|_X - \int_\Omega v_\epsilon(0, x) u_0(x) \, d\mu.
\tag{1.51}
$$

By Hölder's inequality and hypothesis (1.45), we get

$$
\begin{aligned}
\left| \int_\Omega v_\epsilon(0, x) u_0(x) \, d\mu \right| &\leq \left(\int_\Omega v_\epsilon^2(0, x) \, d\mu \right)^{\frac{1}{2}} \left(\int_\Omega u_0^2(x) \, d\mu \right)^{\frac{1}{2}} \\
&\leq C \left(\int_0^T \int_\omega v_\epsilon^2(t, x) \, d\mu \, dt \right)^{\frac{1}{2}} \|u_0\|_X.
\end{aligned}
\tag{1.52}
$$

Thus, by definition of f_ϵ, from (1.51) and (1.52), we get

$$
\begin{aligned}
\|f_\epsilon\|_{L^2(0,T;X)}^2 &= \int_0^T \int_\omega v_\epsilon^2(t, x) \, d\mu \, dt \\
&\leq -\int_0^1 v_\epsilon(0, x) u_0(x) \, d\mu \\
&\leq C \left(\int_0^T \int_\omega v_\epsilon^2(t, x) \, d\mu \, dt \right)^{\frac{1}{2}} \|u_0\|_X.
\end{aligned}
\tag{1.53}
$$

Then we have

$$
\|f_\epsilon\|_{L^2(0,T;X)} = \left(\int_0^T \int_\omega v_\epsilon^2(t, x) \, d\mu \, dt \right)^{\frac{1}{2}} \leq C \|u_0\|_X,
$$

so that $(f_\epsilon)_\epsilon$ is bounded in $L^2(0, T; X)$. Therefore, up to subsequences, we have that $f_\epsilon \rightharpoonup f$ in $L^2(0, T; X)$.

Since u_ϵ solves (1.47), by (1.40) we know that

$$
\begin{aligned}
\|u_\epsilon\|_{C([0,T];X)}^2 + \|u_\epsilon\|_{L^2(0,T;V)}^2 + \|u_\epsilon\|_{H^1(0,T;V^*)}^2 &\leq C \left(\|u_0\|_X^2 + \int_0^T \|f_\epsilon\|_X^2 \, dt \right) \\
&\leq C \|u_0\|_X \leq \tilde{C}
\end{aligned}
$$

for some $\tilde{C} > 0$ and all ϵ. Hence, up to subsequence, recalling the compact embedding $L^2(0, T; X) \cap H^1(0, T; V^*) \hookrightarrow\hookrightarrow C([0, T]; X)$ (see [98]), we get, in particular, that

$$u_\epsilon(T) \to u(T) \text{ in } X \text{ as } \epsilon \to 0. \tag{1.54}$$

Moreover, u is a solution of (1.39), as it is easy to show multiplying the equation in (1.47) by any test function, passing to the limit and using the embedding above together with the fact that $f_\epsilon \rightharpoonup f$ in $L^2(0, T; X)$.

Now, take any $v_T \in X$ and use the solution v of (1.44) as test function in (1.47), and u_ϵ as test function for the problem solved by v; by duality we get

$$\int_\Omega u_\epsilon(T) v_T d\mu = \int_0^T \int_\omega v v_\epsilon d\mu dt + \int_\Omega u_0(x) v(0, x) d\mu$$
$$= -\epsilon \frac{\int_\Omega v_T v_T^\epsilon d\mu}{\|v_T^\epsilon\|_X}$$

by (1.50). Hence, by the Hahn Banach Theorem,

$$\|u_\epsilon(T)\|_X \leq \epsilon.$$

Thus, by (1.54),
$$u(T) = 0 \text{ in } X,$$

as claimed. $\square$

Remark 1.11 We notice that the second part of the proof of the previous theorem provides a constructive way to produce the control function f, passing through some approximating problems. This procedure is strictly related to another notion of controllability, the so-called "approximate controllability" (for instance, see [3, 39, 56, 65, 97]), which we will not treat here.

Remark 1.12 Of course, the equivalence between null controllability and observability holds in more general situations. For instance, see [42, Chap. 2].

Remark 1.13 The importance of Theorem 1.3 is clear: whenever one wants to prove null controllability, it will be enough to prove an observability inequality like (1.45). This is what we shall do in the next chapters.

Chapter 2
The Non Singular Case: $\lambda = 0$

Abstract We show Carleman estimates for parabolic problems in divergence or non divergence form with degeneracy at the boundary or in the interior of the space domain. By them we obtain observability inequalities, proving that the problems are null controllable.

Keywords Carleman estimates · Observability inequalities · Null controllability

In this chapter we will consider the problem

$$\begin{cases} u_t - \mathcal{A}u = f\chi_\omega, & (t, x) \in Q, \\ Bu_{|\partial\Omega} = 0, & t \in (0, T), \\ u(0, x) = u_0(x) \in X, & x \in \Omega, \end{cases} \qquad (2.1)$$

where, as before, B is a boundary operator (typically, Dirichlet or Neumann), χ_ω is the characteristic function of the subdomain $\omega \subset \Omega$, $f \in L^2(0, T; X)$ and

$$\mathcal{A}u = \Delta u, \quad \mathcal{A}u = (a(x)u_x)_x \text{ or } \mathcal{A}u = a(x)u_{xx}.$$

Here a can be strictly positive or can degenerate somewhere.

If a is strictly positive, null controllability has been widely studied in recent years. Indeed, concerning the heat equation, we refer, for example, to [56, 62, 66, 67, 87, 102, 111, 122, 126] and the references therein. In the general case, i.e. $\mathcal{A}u = (a(x)u_x)_x$ or $\mathcal{A}u = a(x)u_{xx}$ and $a(x) > 0$ for all $x \in \bar{\Omega}$, the reader is referred to [61, 87] or [112] for the approach based on Carleman estimates. Several results have also been obtained for semilinear nondegenerate equations, see, in particular, [61, 66, 67, 87, 95]. From now on, we focus on *degenerate* equations.

© The Author(s), under exclusive license to Springer Nature Switzerland AG 2021
G. Fragnelli and D. Mugnai, *Control of Degenerate and Singular Parabolic Equations*, SpringerBriefs in Mathematics, https://doi.org/10.1007/978-3-030-69349-7_2

2.1   The Boundary Degenerate Case

In the following pages we will consider (2.1) when a is 0 at the boundary of the spatial domain. We will see that the first result in this context is negative, in the sense that the classical null controllability may fail. In order to make the reading easier, we divide this section in two subsections: the divergence case and the non divergence one.

2.1.1   The Divergence Case

In [37] the authors considered (2.1) in divergence form, i.e.

$$\begin{cases} u_t - (a(x)u_x)_x = f\chi_\omega, & (t, x) \in Q_T := (0, T) \times (0, 1), \\ Bu_{|\partial\Omega} = 0, & t \in (0, T), \\ u(0, x) = u_0(x) \in X, & x \in (0, 1), \end{cases} \tag{2.2}$$

and $\omega \subset\subset (0, 1)$. They proved that, even if a vanishes only at $x = 0$ (being $\Omega = (0, 1)$), for example $a(x) = x^\alpha$ with $\alpha > 2$, using the following classical change of variables

$$y := \int_x^1 \frac{1}{\sqrt{a(\sigma)}} d\sigma = \frac{2}{2 - \alpha}(1 - x^{1-\alpha/2}), \quad \text{and} \quad U(t, y) := a^{\frac{1}{4}}(x)u(t, x),$$

we get that U solves the equation

$$U_t - U_{yy} + c(y)U = F\chi_{\tilde\omega}, \quad (t, x) \in (0, T) \times (0, +\infty),$$

where

$$c(y) = \frac{\alpha}{4}\left(\frac{3\alpha}{4} - 1\right)\left(1 + \frac{\alpha - 2}{2}y\right)^{-2},$$

$$F(t, y) = f\left(t, \left(1 + \frac{\alpha - 2}{2}y\right)^{\alpha/2(2-\alpha)}\right),$$

and $\tilde\omega \subset\subset (0, +\infty)$ is the transformed control region which is still *bounded*. Then, applying a result of Micu and Zuazua [102], one can conclude that (2.2) is not null controllable. The lack of null controllability comes from the fact that the heat equation holds in an unbounded domain, while the control region is bounded, thus an unbounded region is left without control. With an analogous technique and more sophisticated calculations, one can obtain the same conclusion also when $\alpha = 2$. For this reason, in [28, 32, 33, 37] new notions of null controllability have been developed. For the sake of simplicity, in the rest of this section we assume that a vanishes only at $x = 0$ (the case $a(1) = 0$ can be treated in a similar way). The first notion is the following.

Definition 2.1 (*Regional null controllability* [37]) Let $\omega = (\zeta, \xi) \subset\subset (0, 1)$. Problem (2.2) is said to be *regional null controllable* in time T if for all $u_0 \in L^2(0, 1)$, and $\delta \in (0, \xi - \zeta)$, there exists $f \in L^2(Q_T)$ such that the solution u of (2.2) satisfies

$$u(T, x) = 0 \text{ for } x \in (\zeta + \delta, 1). \tag{2.3}$$

We note that the global null controllability is a strong property in the sense that it is automatically preserved with time. More precisely, if $u(T) \equiv 0$ in $(0, 1)$ and if we stop controlling the system at time T, then for all $t \geq T$, $u(t) \equiv 0$ in $(0, 1)$. On the contrary, regional null controllability is a weaker property: due to the uncontrolled part on $(0, \zeta + \delta)$, (2.3) is no more preserved with time if we stop controlling at time T. Thus, it is important to improve the previous result, as shown in [37], proving that the solution can be forced to vanish identically on $(\zeta + \delta, 1)$ during a given time interval (T, T'), i.e. that the solution is persistent regional null controllable:

Definition 2.2 (*Persistent regional null controllability* [37]) Problem (2.2) is said to be *persistent regional null controllable* in time $T' > T > 0$ if for all $u_0 \in L^2(0, 1)$, and $\delta \in (0, \xi - \zeta)$, there exists $f \in L^2((0, T') \times (0, 1))$ such that the solution u of (2.2) satisfies

$$u(t, x) = 0 \text{ for } (t, x) \in (T, T') \times (\zeta + \delta, 1). \tag{2.4}$$

The previous properties have been proved in the cited papers assuming that

$$a \in C^1((0, 1]), \ a > 0 \text{ in } (0, 1] \text{ and } a(0) = 0.$$

However, while in [37] the proof is based on an observability inequality for a suitable adjoint problem and such an inequality is essentially obtained via Carleman estimates for nondegenerate parabolic operators, in [32, 33] the authors used only a technique based on cut off functions. Clearly, these techniques can be adapted to the case when a degenerates also at $x = 1$.

Obviously, also the persistent regional null controllability is a weaker property with respect to the global null controllability. For this reason, in [36] the authors looked for suitable assumptions on the functions and/or on the variables of the problem in order to obtain (GNC). In particular, they considered the simple equation

$$u_t - (x^\alpha u_x)_x = f(t, x)\chi_{(\zeta, \xi)}(x), \quad (t, x) \in Q_T, \tag{2.5}$$

under suitable boundary conditions and proved that if

$$\alpha \in \left[0, \frac{1}{2}\right] \cup \left[\frac{5}{4}, 2\right),$$

then (2.5) is (GNC) (see [36, Theorem 1.1]).

The restriction on α is due to the proofs of Carleman estimates and observability results for the associated adjoint problem. This restriction is overcome in [1, 101] adding another assumption on the function a:

Hypothesis 2.1 The function $a : [0, 1] \to \mathbb{R}_+$ is **(WD)** or **(SD)**. Moreover, in the strongly degenerate case we assume that

$$\begin{cases} \exists \theta \in (1, K_a] \text{ such that } x \mapsto \dfrac{a(x)}{x^\theta} \text{ is nondecreasing near } 0, & \text{if } K_a > 1, \\[2mm] \exists \theta \in (0, 1) \text{ such that } x \mapsto \dfrac{a(x)}{x^\theta} \text{ is nondecreasing near } 0, & \text{if } K_a = 1. \end{cases}$$

Again, the prototype is $a(x) = x^\alpha$, $\alpha \in (0, 2)$.   □

Now, introduce the following weighted Hilbert spaces:
Case (WD):

$$H_a^1(0, 1) := \{u \in L^2(0, 1) \mid u \text{ absolutely continuous in } [0, 1],$$
$$\sqrt{a}u' \in L^2(0, 1) \text{ and } u(1) = u(0) = 0\}$$

and

$$H_a^2(0, 1) := \{u \in H_a^1(0, 1) \mid au' \in H^1(0, 1)\}.$$

Case (SD):

$$H_a^1(0, 1) := \{u \in L^2(0, 1) \mid u \text{ locally absolutely continuous in } (0, 1],$$
$$\sqrt{a}u' \in L^2(0, 1) \text{ and } u(1) = 0\}$$

and

$$H_a^2(0, 1) := \{u \in H_a^1(0, 1) \mid au' \in H^1(0, 1)\}$$
$$= \{u \in L^2(0, 1) \mid u \text{ locally absolutely continuous in } (0, 1],$$
$$au \in H_0^1(0, 1), au' \in H^1(0, 1) \text{ and } (au')(0) = 0\},$$

with the norms

$$\|u\|_{H_a^1}^2 := \|u\|_{L^2(0,1)}^2 + \|\sqrt{a}u'\|_{L^2(0,1)}^2, \quad \text{and} \quad \|u\|_{H_a^2}^2 := \|u\|_{H_a^1}^2 + \|(au')'\|_{L^2(0,1)}^2.$$

Remark 2.1 Notice the improved regularity used here in the **(WD)** case, with respect to the spaces $\mathcal{H}_a^1(0, 1)$ in Definition 1.3.

To both types of degeneracy we associate its own boundary condition at $x = 0$. Hence, given $f \in L^2(Q_T)$, we consider:

$$\begin{cases} u_t - (au_x)_x = f(t, x)\chi_\omega(x), & (t, x) \in Q_T, \\ u(t, 1) = 0, & t \in (0, T), \\ \begin{cases} u(t, 0) = 0, & \text{for } (\mathbf{WD}) \\ \text{or} \\ (au_x)(t, 0) = 0, & \text{for } (\mathbf{SD}) \end{cases}, & t \in (0, T), \\ u(0, x) = u_0(x), & x \in (0, 1). \end{cases} \quad (2.6)$$

As usual, the subdomain $\omega = (\zeta, \xi) \subset\subset (0, 1)$ is the control region.

The previous problem is well posed in the following sense:

Theorem 2.1 (Theorem 2.6 [1]) *Assume Hypothesis 2.1. Then, for all $f \in L^2(Q_T)$ and $u_0 \in L^2(0, 1)$, there exists a unique weak solution*

$$u \in C^0([0, T]; L^2(0, 1)) \cap L^2(0, T; H^1_a(0, 1))$$

of (2.6). Moreover, if $u_0 \in H^1_a(0, 1)$, then

$$u \in H^1(0, T; L^2(0, 1)) \cap L^2(0, T; H^2_a(0, 1)) \cap C([0, T]; H^1_a(0, 1)),$$

and there exists a positive constant C such that

$$\sup_{t \in [0, T]} \left(\|u(t)\|^2_{H^1_a} \right) + \int_0^T \left(\|u_t\|^2_{L^2(0, 1)} + \|(au_x)_x\|^2_{L^2(0, 1)} \right) dt$$
$$\leq C \left(\|u_0\|^2_{H^1_a} + \|f\|^2_{L^2(Q_T)} \right).$$

Of course, in view of Definition 1.4, since

$$H^1(0, T; L^2(0, 1)) \subset H^1(0, T; H^1_a(0, 1)^*),$$

we have that, if $u_0 \in L^2(0, 1)$ and $f \in L^2(Q_T)$, a function

$$u \in C([0, T]; L^2(0, 1)) \cap L^2(0, T; H^1_a(0, 1))$$

is a (weak) solution of (2.6) if

$$\int_0^1 u(T, x)\varphi(T, x)\, dx - \int_0^1 u_0(x)\varphi(0, x)\, dx - \int_0^T \int_0^1 u\varphi_t \, dx dt =$$
$$-\int_0^T \int_0^1 au_x\varphi_x \, dx dt + \int_0^T \int_0^1 f\varphi\chi_\omega \, dx dt$$

for all $\varphi \in H^1(0, T; L^2(0, 1)) \cap L^2(0, T; H^1_a(0, 1))$.

The proof of Theorem 2.1 is based on classical tools of semigroup theory, thanks to the following result, see [36]:

Proposition 2.1 *If* $\mathcal{A}u := (au')'$ *with domain* $D(\mathcal{A}) = H_a^2(0, 1)$, *the operator* $\mathcal{A}$: $D(\mathcal{A}) \to L^2(0, 1)$ *is closed, self-adjoint and negative with dense domain.*

Observe that the well posedness of (2.6) is independent of the degree of degeneracy of a. Actually, the (SD) case is, as we have seen before, the interesting case from the viewpoint of null controllability since (2.1) when $a(x) = x^\alpha$, with $\alpha > 2$, can fail to be null controllable on the whole interval $[0, 1]$, and regional null controllability is the only property that can be expected, see [37].

In order to prove null controllability, in [1] the authors used the equivalence result stated in Theorem 1.3. Hence, the crucial key is to obtain the observability inequality (1.45). The idea is to use Carleman estimates for the non homogeneous associated adjoint problem (1.44) that in this case reads

$$
\begin{cases}
v_t + (a(x)v_x)_x = h, & (t, x) \in Q_T, \\
v(t, 1) = 0, & t \in (0, T), \\
\begin{cases}
v(t, 0) = 0, & \text{for (\textbf{WD})} \\
\text{or} \\
(av_x)(t, 0) = 0, & \text{for (\textbf{SD})}
\end{cases} & , \quad t \in (0, T).
\end{cases}
\tag{2.7}
$$

To obtain Carleman estimates we define the following weight function that will be used also later:

$$
\varphi(t, x) := \Theta(t)\psi(x),
\tag{2.8}
$$

where

$$
\Theta(t) := \frac{1}{[t(T - t)]^4},
\tag{2.9}
$$

and

$$
\psi(x) := c_1 \left(\int_0^x \frac{y}{a(y)} dy - c_2 \right),
\tag{2.10}
$$

with $c_2 > \frac{1}{a(1)(2-K)}$ and $c_1 > 0$.

Remark 2.2 Usually, for classical parabolic equations, it is enough to choose $\Theta(t) := \frac{1}{[t(T-t)]^k}$ with $k \geq 1$. Here we need $k \geq 4$ (and so it is natural to choose $k = 4$), since, at a certain point, one needs to estimate $\ddot{\Theta}$ with $\Theta^{3/2}$ (see Sect. 2.3).

Using Lemma 1.1 we can prove as in [1] that the next estimate holds:

Theorem 2.2 (Theorem 3.1 [1]) *Assume Hypothesis 2.1 and let* $T > 0$ *be given. Then there exist two positive constants* C *and* s_0, *such that every solution* v *of* (2.7) *belonging to* $L^2(0, T; H_a^2(0, 1)) \cap H^1(0, T; H_a^1(0, 1))$ *satisfies, for all* $s \geq s_0$,

$$\int_0^T \int_0^1 \left(s\Theta a(x)v_x^2 + s^3\Theta^3 \frac{x^2}{a(x)} v^2 \right) e^{2s\varphi(t,x)} dxdt$$

$$\leq C \left(\int_0^T \int_0^1 h^2 e^{2s\varphi(t,x)} dxdt + sa(1) \int_0^T \Theta e^{2s\varphi(t,1)} v_x^2(t,1) dt \right). \tag{2.11}$$

Here and in the following, by u_x^2 we mean $(u_x)^2$.

The proof of the previous theorem is given in the last section of this chapter, in order to give an idea on how proving Carleman estimate. Indeed, such inequalities are in general quite complicated, so that we will only prove this one. For the other proofs we shall refer to the related papers.

An estimate similar to (2.11) was also proved in [101] under slightly different assumptions (see also the considerations after Proposition 1.1).

As a consequence of (2.11), we derive the observability inequality (1.45) for any solution of the adjoint problem (2.7). We conclude this subsection with the proof of such a result, but we underline the fact that, with suitable changes, it holds in the more general case $\lambda \neq 0$. Then, thanks to Theorem 1.3 the null controllability of (2.6) follows.

The first step to obtain (1.45) is to prove the next lemma.

Lemma 2.1 (Lemma 4.2 [1]) *Let* $\omega = (\zeta, \xi) \subset\subset (0, 1)$. *Then there exist two positive constants* C *and* s_0 *such that every solution* $v \in \mathcal{W} := C^1([0, T]; L^2(0, 1)) \cap C([0, T]; D(\mathcal{A}))$ *of*

$$\begin{cases} v_t + (av_x)_x = 0, & (t, x) \in Q_T, \\ v(t, 1) = 0, & t \in (0, T), \\ \begin{cases} v(t, 0) = 0, & \text{for (WD)}, \\ \text{or} & t \in (0, T), \\ (av_x)(t, 0) = 0, & \text{for (SD)}, \end{cases} \\ v(T, x) = v_T(x) \in D(\mathcal{A}^2) \end{cases} \tag{2.12}$$

satisfies, for all $s \geq s_0$,

$$\int_0^T \int_0^1 \left(s\Theta a v_x^2 + s^3\Theta^3 \frac{x^2}{a} v^2 \right) e^{2s\varphi} dxdt \leq C \int_0^T \int_\omega v^2 dxdt.$$

Here $\mathcal{A}u := (au_x)_x$, $D(\mathcal{A}) := H_a^2(0, 1)$ *and*

$$D(\mathcal{A}^2) = \left\{ u \in D(\mathcal{A}) \,\middle|\, \mathcal{A}u \in D(\mathcal{A}) \right\}.$$

Observe that $D(\mathcal{A}^2)$ is densely defined in $D(\mathcal{A})$ (see, for example, [22, Lemma 7.2]) and hence in $L^2(0, 1)$. Obviously (see, for example, [22, Theorem 7.5])

$$\mathcal{W} \subset C^1\big([0, T]; H_a^2(0, 1)\big) \subset L^2\big(0, T; H_a^2(0, 1)\big) \cap H^1\big(0, T; H_a^1(0, 1)\big)$$
$$\subset C\big([0, T]; L^2(0, 1)\big) \cap L^2\big(0, T; H_a^1(0, 1)\big).$$

The proof of the previous result is based on the next two results.

Lemma 2.2 (Lemma 4.3 [1]) (Caccioppoli's inequality) *Let* $\omega' \subset\subset \omega \subset\subset (0, 1)$ *and* $v \in \mathcal{W}$ *solution of (2.12). Then there exists a positive constant* C *such that for any* $s > 0$

$$\int_0^T \int_{\omega'} v_x^2 e^{2s\varphi} dx dt \leq C \int_0^T \int_\omega v^2 dx dt.$$

Proposition 2.2 (Proposition 4.4 [1] or Theorem 3.1 [78]) (Classical Carleman estimates) *Consider the problem*

$$\begin{cases} z_t + (az_x)_x = h \in L^2(Q_T), \\ z(t, 0) = z(t, 1) = 0, \ t \in (0, T), \end{cases} \tag{2.13}$$

where $a \in C^1\big([0, 1]\big)$ *is a strictly positive function. Then for every* $r > 0$ *there exists* $s_0 = s_0(r) > 0$ *such that for any* $s > s_0$, *any solution* z *of (2.13) satisfies*

$$\int_0^T \int_0^1 s^3 \Theta^3 e^{3rY(x)} z^2 e^{-2s\Phi(t,x)} dx dt + \int_0^T \int_0^1 s\Theta e^{rY(x)} z_x^2 e^{-2s\Phi(t,x)} dx dt$$
$$\leq c \int_0^T \int_0^1 e^{-2s\Phi(t,x)} h^2 dx dt - c \int_0^T \big[\sigma(t, \cdot) e^{-2s\Phi(t,\cdot)} a(\cdot) z_x^2\big]_{x=0}^{x=1} dt, \tag{2.14}$$

for some positive constant c. *Here the functions* Y, σ *and* Φ *are defined in the following way:*

$$Y(x) := \int_x^1 \frac{1}{\sqrt{a(y)}} dy, \qquad \sigma(t, x) := rs\Theta(t) e^{rY(x)}$$

and

$$\Phi(t, x) := \Theta(t)\Psi(x),$$

where $(t, x) \in Q_T, r, s > 0$ *and* $\Psi(x) := \big(e^{2rY(0)} - e^{rY(x)}\big).$

(Observe that $\Phi > 0$ and $\Phi(t, x) \to +\infty$, as $t \downarrow 0, t \uparrow T$.)

Proof (*Proof of Lemma 2.1*) Recalling that $\omega = (\zeta, \xi)$, let us consider a smooth cut-off function $\rho : \mathbb{R} \to \mathbb{R}$, such that

$$\begin{cases} 0 \leq \rho(x) \leq 1, & \text{for all } x \in \mathbb{R}, \\ \rho(x) = 1, & x \in \left(0, \frac{2\zeta + \xi}{3}\right), \\ \rho(x) = 0, & x \in \left(\frac{\zeta + 2\xi}{3}, 1\right), \end{cases}$$

and $\rho(x) > 0$ for $x < (\zeta + 2\xi)/3$. We define $w := \rho v$ where v solves (2.12). Then w satisfies

$$\begin{cases} w_t + (aw_x)_x = (a\rho_x v)_x + \rho_x a v_x =: \tilde{h}, & (t,x) \in Q_T, \\ w(t,1) = 0, & t \in (0,T), \\ \left\{ \begin{array}{ll} w(t,0) = 0, & \text{for (\textbf{WD}),} \\ \text{or} & t \in (0,T). \\ (aw_x)(t,0) = 0, & \text{for (\textbf{SD}),} \end{array} \right. \end{cases} \qquad (2.15)$$

Therefore, applying Theorem 2.2 and using the definition of w, we have

$$\int_0^T \int_0^1 \left(s\Theta a(x)w_x^2 + s^3\Theta^3 \frac{x^2}{a(x)}w^2 \right)e^{2s\varphi}\,dxdt \leq C \int_0^T \int_0^1 e^{2s\varphi}\tilde{h}^2 dxdt$$
$$+ \int_0^T sa(1)\Theta(t)w_x^2(t,1)e^{2s\varphi(t,1)}\,dt \leq C \int_0^T \int_{\omega'} e^{2s\varphi}(v_x^2 + v^2)\,dxdt,$$

where $\omega' := ((2\zeta + \xi)/3), (\zeta + 2\xi)/3)$ and s larger than a certain s_0. Thanks to Lemma 2.2, we can estimate the first term in the last integral using a constant which is independent of $s \geq s_0$, obtaining

$$\int_0^T \int_0^1 \left(s\Theta a(x)w_x^2 + s^3\Theta^3 \frac{x^2}{a(x)}w^2 \right)e^{2s\varphi}\,dxdt \leq C \int_0^T \int_\omega v^2 dxdt \qquad (2.16)$$

for every $s \geq s_0$. On $(\zeta,1)$ the equation is uniformly parabolic, so we can use the classical Carleman estimate (2.14) obtaining the same bound for v.

Now define $z := \eta v$, where $\eta := 1 - \rho$. Then z satisfies (2.13), with $\bar{h} := (a\eta_x v)_x + \eta_x a v_x$, and (2.14) holds, as well. By definitions of Y and Φ, and using again Lemma 2.2, we get

$$\int_0^T \int_0^1 s^3\Theta(t)^3 e^{3rY(x)}z^2(t,x)e^{-2s\Phi(t,x)}\,dxdt$$
$$+ \int_0^T \int_0^1 s\Theta(t)e^{rY(x)}z_x^2(t,x)e^{-2s\Phi(t,x)}\,dxdt \qquad (2.17)$$
$$\leq c \int_0^T \int_0^1 e^{-2s\Phi(t,x)}\bar{h}^2 dxdt \leq C \int_0^T \int_\omega v^2 dxdt.$$

Since $v = w + z$ (recall that $w = \rho v$ and $z = \eta v$), then $v^2 \leq 2(w^2 + z^2)$ and $v_x^2 \leq 2(w_x^2 + z_x^2)$. Thus by (2.16), one has

$$\int_0^T \int_0^1 \left(s\Theta a(x)v_x^2 + s^3\Theta^3 \frac{x^2}{a(x)}v^2\right)e^{2s\varphi}\,dx\,dt$$

$$\leq 2\int_0^T \int_0^1 \left(s\Theta a(x)w_x^2 + s^3\Theta^3 \frac{x^2}{a(x)}w^2\right)e^{2s\varphi}\,dx\,dt$$

$$+ 2\int_0^T \int_0^1 \left(s\Theta a(x)z_x^2 + s^3\Theta^3 \frac{x^2}{a(x)}z^2\right)e^{2s\varphi}\,dx\,dt$$

$$\leq C\int_0^T \int_\omega v^2\,dx\,dt + 2\int_0^T \int_0^1 \left(s\Theta(t)a(x)z_x^2 + s^3\Theta(t)^3 \frac{x^2}{a(x)}z^2\right)e^{2s\varphi(t,x)}\,dx\,dt.$$

By definition of φ and choosing c_1 in (2.10) so that

$$c_1 \geq \frac{(2-K_a)a(1)}{a(1)c_2(2-K_a)-1}(e^{2rY(0)} - 1),$$

one can prove that there exists a positive constant k (for example $k = \min\left\{\max_{[\zeta,1]} a(x), \frac{1}{\min_{[\zeta,1]} a(x)}\right\}$) such that

$$a(x)e^{2s\varphi(t,x)} \leq ke^{rY(x)}e^{-2s\Phi(t,x)}$$

and

$$\frac{x^2}{a(x)}e^{2s\varphi(t,x)} \leq ke^{rY(x)}e^{-2s\Phi(t,x)} \leq ke^{3rY(x)}e^{-2s\Phi(t,x)},$$

$\forall\,(t,x) \in [0,T] \times [\zeta,1]$. By using (2.17) and the previous estimates, we find

$$\int_0^T \int_0^1 \left(s\Theta(t)a(x)z_x^2 + s^3\Theta(t)^3 \frac{x^2}{a(x)}z^2\right)e^{2s\varphi(t,x)}\,dx\,dt$$

$$= \int_0^T \int_\zeta^1 \left(s\Theta(t)a(x)z_x^2 + s^3\Theta(t)^3 \frac{x^2}{a(x)}z^2\right)e^{2s\varphi(t,x)}\,dx\,dt$$

$$\leq k\int_0^T \int_\zeta^1 \left(s\Theta(t)z_x^2 + s^3\Theta(t)^3 \frac{x^2}{a(x)}z^2\right)e^{3rY(x)}e^{-2s\Phi(t,x)}\,dx\,dt$$

$$\leq kC\int_0^T \int_\omega v^2\,dx\,dt.$$

Thus

$$\int_0^T \int_0^1 \left(s\Theta a(x)v_x^2 + s^3\Theta^3 \frac{x^2}{a(x)}v^2\right)e^{2s\varphi}\,dx\,dt \leq C\int_0^T \int_\omega v^2\,dx\,dt,$$

for some positive constant C.   □

As a consequence, one can prove the observability inequality

Proposition 2.3 *Assume that Hypothesis 2.1 is satisfied and let $T > 0$ be given. Then there exists a positive constant C such that every solution $v \in \mathcal{W}$ of (2.12) satisfies*

$$\int_0^1 v^2(0, x)dx \le C_T \int_0^T \int_\omega v^2(t, x)dxdt. \tag{2.18}$$

Proof Notice that, being the final datum more regular, $v \in \mathcal{W}$ satisfies (2.12) point-wise. Hence, multiplying the equation $v_t + (av_x)_x = 0$ by v_t and integrating by parts over $(0, 1)$, one has

$$
\begin{aligned}
0 &= \int_0^1 (v_t + (av_x)_x)v_t dx = \int_0^1 (v_t^2 + (av_x)_x v_t)dx \\
&= \int_0^1 v_t^2 dx + [av_x v_t]_{x=0}^{x=1} - \int_0^1 av_x v_{tx} dx \\
&\ge \int_0^1 v_t^2 dx + [av_x v_t]_{x=0}^{x=1} - \frac{1}{2}\frac{d}{dt}\int_0^1 av_x^2 dx \ge -\frac{1}{2}\frac{d}{dt}\int_0^1 av_x^2 dx.
\end{aligned}
$$

Thus, the function $t \mapsto \int_0^1 a(v_x)^2 dx$ is increasing for all $t \in [0, T]$. In particular,

$$\int_0^1 av_x(0, x)^2 dx \le \int_0^1 av_x(t, x)^2 dx$$

for every $t \in [0, T]$. Integrating the last inequality over $\left[\dfrac{T}{4}, \dfrac{3T}{4}\right]$, Θ being bounded therein, we find

$$
\begin{aligned}
\int_0^1 a(v_x)^2(0, x)dx &\le \frac{2}{T}\int_{\frac{T}{4}}^{\frac{3T}{4}}\int_0^1 a(v_x)^2(t, x)dxdt \\
&\le C_T \int_{\frac{T}{4}}^{\frac{3T}{4}}\int_0^1 s\Theta a(v_x)^2(t, x)e^{2s\varphi}dxdt.
\end{aligned}
$$

Hence, by Lemma 2.1 and the previous inequality, there exists a positive constant C such that

$$\int_0^1 a(v_x)^2(0, x)dx \le C \int_0^T \int_\omega v^2 dxdt. \tag{2.19}$$

Applying the Hardy–Poincaré inequality given in Proposition 1.1, we get

$$\int_0^1 \frac{a(x)}{x^2}v(0, x)^2 dx \le C \int_0^1 av_x^2(0, x)dx \le C \int_0^T \int_\omega v^2 dxdt.$$

But, since the map $x \mapsto \dfrac{a(x)}{x^2}$ is nonincreasing, as the inequality $xa'(x) \leq K_a a(x)$ easily implies, then $\dfrac{a(x)}{x^2} \geq a(1) > 0$ and so

$$\int_0^1 v(0, x)^2 dx \leq C \int_0^T \int_\omega v^2 dx dt$$

for some universal constant $C > 0$.   □

We are now ready for the validity of the observability inequality for solutions of (2.7).

Proposition 2.4 *Assume Hypothesis 2.1. Then there exists a positive constant C such that every solution $v \in C([0, T]; L^2(0, 1)) \cap L^2(0, T; H_a^1(0, 1))$ of (2.7) satisfies (1.45).*

Proof The proof is now standard, but we give it with some precise references: let $v_T \in L^2(0, 1)$ and let v be the solution of (2.7) associated to v_T. Since $D(\mathcal{A}^2)$ is densely defined in $L^2(0, 1)$, there exists a sequence $(v_T^n)_n \subset D(\mathcal{A}^2)$ which converges to v_T in $L^2(0, 1)$. Now, consider the solution v_n associated to v_T^n.

By [1, Proposition 2.5], $\mathcal{A}$ is closed, thus, by [51, Theorem II.6.7], we get that $(v_n)_n$ converges to a certain v in $C([0, T]; L^2(0, 1))$, so that

$$\lim_{n \to +\infty} \int_0^1 v_n^2(0, x) dx = \int_0^1 v^2(0, x) dx,$$

and also

$$\lim_{n \to +\infty} \int_0^T \int_\omega v_n^2 dx dt = \int_0^T \int_\omega v^2 dx dt.$$

But, by Proposition 2.3 we know that

$$\int_0^1 v_n^2(0, x) dx \leq C_T \int_0^T \int_\omega v_n^2 dx dt.$$

Thus the thesis follows.   □

2.1.2 The Non Divergence Case

The null controllability can fail also in the non divergence case. Indeed, as before, we can take $a(x) = x^\alpha$ with $\alpha > 2$. By introducing the new variables (see [29, Remark 4.6])

$$y := \int_x^1 \frac{d\sigma}{\sqrt{a(\sigma)}} = \frac{2}{2-\alpha}(1 - x^{1-\alpha/2}), \quad U(t, y) := a(x)^{-1/4}u(t, x),$$

the degenerate equation

$$u_t - a(x)u_{xx} = f\chi_\omega$$

set in $(0, 1)$ with $\omega \subset\subset (0, 1)$ becomes

$$U_t - U_{yy} + c(y)U = F\chi_{\tilde{\omega}}, \tag{2.20}$$

for a suitable control function F, with $y \in (0, +\infty)$, $\tilde{\omega} \subset\subset (0, +\infty)$ and c given by

$$c(y) = \frac{\alpha}{4}\left(\frac{3\alpha}{4} - 1\right)\left(1 + \left(\frac{\alpha}{2} - 1\right)y\right)^{-2}.$$

Thus the degenerate heat equation in $(0, 1)$ is transformed into a non degenerate heat equation in the unbounded domain $(0, +\infty)$ with a regular potential term. Using a result of Escauriaza, Seregin and Sverak (see [54] or [55]), that generalizes the work of Micu and Zuazua (see [102]), we deduce, as for the divergence case, that (2.1) is not null controllable. Also in this case we can obtain a regional or persistent regional null controllability. Obviously, the most interesting result is the global null controllability, on which we will put our attention.

So, let us introduce the degenerate problem in non divergence form we shall treat, namely

$$\begin{cases} u_t - a(x)u_{xx} = f(t, x)\chi_\omega(x), & (t, x) \in Q_T, \\ u(t, 0) = u(t, 1) = 0, & t \in (0, T), \\ u(0, x) = u_0(x), & x \in (0, 1). \end{cases} \tag{2.21}$$

As for the previous subsection, we introduce, first of all, the Hilbert spaces where the problem will be considered. Actually, one can think to deduce well posedness and null controllability for (2.21) from the ones of (2.6). But it is not true. Indeed, the equation in (2.21) can be recast in divergence form as

$$u_t - (a(x)u_x)_x + a'(x)u_x = f(t, x)\chi_\omega(x) \tag{2.22}$$

at the price of adding the drift term $a'u_x$. Such an addition has major consequences. Indeed, as described in [7], degenerate equations of the form (2.22), are well posed in $L^2(0, 1)$ under the structural assumption

$$|a'(x)| \leq C\sqrt{a(x)} \tag{2.23}$$

for some positive constant C. Now, imposing (2.23) for $a(x) = x^\alpha$ gives $\alpha \geq 2$. So, in view of the above considerations, the conditions that ensure that (2.22) is well posed prevent (2.22) from being null-controllable.

Thus, in [29, 30], the null controllability for (2.21) is studied considering the following Hilbert spaces (we refer to [70] for Neumann boundary condition):

$$L^2_{\frac{1}{a}}(0,1) := \left\{ u \in L^2(0,1) \,\Big|\, \int_0^1 u^2 \frac{1}{a} dx < \infty \right\},$$

$$H^1_{\frac{1}{a}}(0,1) := L^2_{\frac{1}{a}}(0,1) \cap H^1_0(0,1)$$

and

$$H^2_{\frac{1}{a}}(0,1) := \left\{ u \in H^1_{\frac{1}{a}}(0,1) \,\Big|\, au'' \in L^2_{\frac{1}{a}}(0,1) \right\},$$

with the following norms

$$\|u\|^2_{1,\frac{1}{a}} := \int_0^1 u^2 \frac{1}{a} dx + \int_0^1 (u')^2 dx \quad \text{and} \quad \|u\|^2_{2,\frac{1}{a}} := \|u\|^2_{1,\frac{1}{a}} + \int_0^1 a(u'')^2 dx.$$

For simplicity, here we assume that a degenerates only at 0, but we underline that in [29, 30, 70] the function a degenerates at both boundary points.

Theorem 2.3 (Theorem 2.4 [29]; Theorem 2.2 [71]) *Assume that a is* **(WD)** *or* **(SD)**. *Then, for all $f \in L^2(Q_T)$ and $u_0 \in L^2_{\frac{1}{a}}(0,1)$, there exists a unique weak solution $u \in C^0\left([0,T]; L^2_{\frac{1}{a}}(0,1)\right) \cap L^2\left(0,T; H^1_{\frac{1}{a}}(0,1)\right)$ of (2.21). Moreover, if $u_0 \in H^1_{\frac{1}{a}}(0,1)$, then*

$$u \in H^1(0,T; L^2_{\frac{1}{a}}(0,1)) \cap L^2(0,T; H^2_{\frac{1}{a}}(0,1)) \cap C^0([0,T]; H^1_{\frac{1}{a}}(0,1)),$$

and there exists a positive constant C such that

$$\sup_{t \in [0,T]} \|u(t)\|^2_{L^2_{\frac{1}{a}}(0,1)} + \int_0^T \|u\|^2_{H^1_{\frac{1}{a}}(0,1)} dt \leq C \left(\|u_0\|^2_{L^2_{\frac{1}{a}}(0,1)} + \int_0^T \|f\|^2_{L^2_{\frac{1}{a}}(\omega)} dt \right),$$

for a positive constant C.

More generally, on a we assume:

Hypothesis 2.2 The function $a \in C^0[0,1] \cap C^3(0,1]$ is such that $a(0) = 0$, $a > 0$ on $(0,1]$, the function $\dfrac{xa'}{a} \in L^\infty(0,1)$ and there exist $K_a \in (0,2)$ and $C > 0$ such that

$$\frac{xa'(x)}{a(x)} \leq K_a \quad \text{and} \quad \left| \left(\frac{xa'(x)}{a(x)} \right)'' \right| \leq C \frac{1}{a(x)} \quad \forall x \in (0,1). \qquad \square$$

Obviously if $K_a < 1$ then a is weakly degenerate, otherwise it is strongly degenerate. The previous assumption has been improved recently in [74], where, in a more general setting, the function a is required to be only of class $C^2((0,1])$.

Also for the non divergence case, in order to obtain null controllability, we will consider the observability inequality of the associated adjoint problem of (2.21) via

Carleman estimates. Hence, we consider the non homogeneous parabolic problem:

$$\begin{cases} v_t + a(x)v_{xx} = h(t, x), & (t, x) \in Q_T, \\ v(t, 0) = v(t, 1) = 0, & t \in (0, T), \end{cases} \tag{2.24}$$

for all $t \in [0, T]$.

Define

$$\gamma(t, x) := \Theta(t)\psi(x), \tag{2.25}$$

where

$$\psi(x) := (p(x) - 2\|p\|_{L^\infty(0,1)}), \quad p(x) := \int_0^x \frac{y}{a(y)} dy, \tag{2.26}$$

and Θ is as in (2.9).[1] For any (sufficiently smooth) solution v of (2.24), the following Carleman estimate holds:

Theorem 2.4 (Theorem 3.1 [29]) *Assume that Hypothesis 2.2 is satisfied and let $T > 0$ be given. Then there exist two positive constants C and s_0 such that every solution v of (2.24) in $L^2\left(0, T; H^2_{\frac{1}{a}}(0, 1)\right) \cap H^1\left(0, T; H^1_{\frac{1}{a}}(0, 1)\right)$ satisfies, for all $s \geq s_0$,*

$$\int_0^T \int_0^1 \left(s\Theta v_x^2 + s^3\Theta^3 \left(\frac{x}{a}\right)^2 v^2 \right) e^{2s\gamma} dx dt$$

$$\leq C \int_0^T \int_0^1 h^2 \frac{e^{2s\gamma}}{a} dx dt + s\, C \int_0^T \Theta(t)\left[xv_x^2 e^{2s\gamma}\right](t, 1) dt.$$

Thanks to Theorem 2.4, we derive (1.45); hence, the null controllability (1.42) with (1.43) holds true.

2.2 The Degenerate Case in the Interior of the Domain

In a lot of applications the degeneracy occurs at the interior of the space domain, see, for example, the Grushin model [9] or the population model [17, 72–74]. Thus, recently, a lot of interest is given to study the null controllability for this kind of system. If the degenerate coefficient a is regular we can refer to [77] in the divergence form with Dirichlet conditions or to [20] in the non divergence case with Dirichlet or Neumann ones. In the non smooth case we refer to [78]. For simplicity, here we consider only the smooth case.

---

[1] Actually, in [29] much more general assumptions are assumed; precisely, the condition $xa'(x) \leq K_a a(x)$ is assumed only in a left neighborhood of $x = 0$. Thus, one chooses $R > 0$ such that $2 - \frac{xa'(x)}{a(x)} + 4Rx^2 \geq 2 - K_a$ and takes $p(x) := \int_0^x \frac{y}{a(y)} e^{Ry^2} dy$, see [29, p. 601].

As for the case of a boundary degeneracy, also in this situation null controllability results follow from suitable observability inequalities for the associated homogeneous adjoint problems, by applying Theorem 1.3. However, similarly to the previous case, we will not state the controllability results in the general case (for which we refer to the related theorems in the underlying references), but we quote a controllability result, valid in the **(WD)** case, taken from [26], see Theorem 2.6 below.

2.2.1 The Divergence Case

If a degenerates in the interior of the domain, we will consider in the weakly degenerate case the spaces $H_a^1(0, 1)$ and $H_a^2(0, 1)$ defined as before but we will write $H_{a,x_0}^1(0, 1)$ and $H_{a,x_0}^2(0, 1)$ in order to recall that in this case a degenerates at x_0. On the other hand, in the strongly degenerate case we will consider

$$H_{a,x_0}^1(0, 1) := \{u \in L^2(0, 1) : u \text{ is locally absolutely continuous in}$$
$$[0, x_0) \cup (x_0, 1], \sqrt{a}u' \in L^2(0, 1) \text{ and } u(0) = u(1) = 0\}$$

and

$$H_{a,x_0}^2(0, 1) := \{u \in H_{a,x_0}^1(0, 1) | au' \in H^1(0, 1)\},$$

with the norms

$$\|u\|_{H_{a,x_0}^1(0,1)}^2 := \|u\|_{L^2(0,1)}^2 + \|\sqrt{a}u'\|_{L^2(0,1)}^2,$$

and

$$\|u\|_{H_{a,x_0}^2(0,1)}^2 := \|u\|_{H_{a,x_0}^1(0,1)}^2 + \|(au')'\|_{L^2(0,1)}^2.$$

Without difference on the type of degeneration and without other assumptions on the degenerate function, using the previous spaces, one can prove that Theorem 2.1 still holds for

$$\begin{cases} u_t - Au = f\chi_\omega, & (t, x) \in Q_T, \\ u(t, 0) = u(t, 1) = 0, & t \in (0, T), \\ u(0, x) = u_0(x), & x \in (0, 1), \end{cases} \qquad (2.27)$$

(see [78, Theorem 2.1]).

On the contrary, in order to prove null controllability for (2.27), we need an additional assumption. In particular, Hypothesis 2.1 becomes

Hypothesis 2.3 The function $a : [0, 1] \to \mathbb{R}_+$ is **(WD)** or **(SD)**. Moreover, if $K_a > \dfrac{4}{3}$, we suppose that there exists a constant $\vartheta \in (0, K_a]$ such that the function

$$x \mapsto \frac{a(x)}{|x - x_0|^\vartheta} \begin{cases} \text{is nonincreasing on the left of } x = x_0, \\ \text{is nondecreasing on the right of } x = x_0. \end{cases} \qquad (2.28)$$

On the control set $\omega \subset (0, 1)$ we assume:

Hypothesis 2.4   The subset ω is such that

(i) it is an interval which contains the degeneracy point:

$$\omega = (\zeta, \xi) \subset (0, 1) \text{ is such that } x_0 \in \omega, \qquad (2.29)$$

or

(ii)

$$\omega = \omega_1 \cup \omega_2, \qquad (2.30)$$

where

$$\omega_i = (\zeta_i, \xi_i) \subset (0, 1), \ i = 1, 2, \ \text{and} \ \xi_1 < x_0 < \zeta_2.$$

Remark 2.3   Condition (2.28) is more general than the corresponding one for $x_0 = 0$ required in Hypothesis 2.1 for the **(SD)** case. Indeed, in this case we require it only in the sub-case $K_a > \frac{4}{3}$ of the **(SD)** case. On the other hand, let us note that requiring (2.28), also with $x_0 = 0$, together with the condition given in Definition 1.2, implies $\vartheta a(x) \le (x - x_0)a'(x) \le K_a a(x)$ in $(0, 1)$, so that a' is automatically bounded away from 0 far from x_0.

Remark 2.4   Actually, in [78] the case of a not of class C^1 is considered. More precisely, it is assumed to be of class $W^{1,1}$ in the **(WD)** case and of class $W^{1,\infty}$ in the **(SD)** case. This lack of regularity implies some additional requirements. In particular, Hypothesis 2.3 must be completed requiring that, when $K_a > \frac{3}{2}$, the map in (2.28) is bounded below away from 0 and there exists a constant $\Sigma > 0$ such that

$$|a'(x)| \le \Sigma |x - x_0|^{2\vartheta - 3}, \quad x \in [0, 1]. \qquad (2.31)$$

These additional requirements for the sub-case $K_a > 3/2$ are technical ones, and are used just to guarantee the convergence of some integrals (see [78, Appendix]). Of course, the prototype $a(x) = |x - x_0|^{K_a}$ satisfies such conditions with $\vartheta = K_a$.

As in the previous section, the crucial key to prove Carleman estimate is to introduce a good weight. Also in this case, we consider the function φ given in (2.8), where ψ is defined in this way

$$\psi(x) := c_1 \left[\int_{x_0}^x \frac{y - x_0}{a(y)} dy - c_2 \right], \qquad (2.32)$$

with $c_2 > \max \left\{ \dfrac{(1 - x_0)^2}{a(1)(2 - K_a)}, \dfrac{x_0^2}{a(0)(2 - K_a)} \right\}$ and $c_1 > 0$.

Then, the following estimate holds

Theorem 2.5 (Theorem 4.1 [78]) *Assume Hypothesis 2.3 and let $T > 0$. Then there exist two positive constants C and s_0, such that every solution $v \in L^2(0, T; H^2_{a,x_0}$ $(0, 1)) \cap H^1(0, T; H^1_{a,x_0}(0, 1))$ of*

$$\begin{cases} v_t + (a(x)v_x)_x = h, & (t, x) \in Q_T, \\ v(t, 1) = v(t, 0) = 0, & t \in (0, T), \end{cases} \qquad (2.33)$$

satisfies, for all $s \geq s_0$,

$$\int_0^T \int_0^1 \left(s\Theta a(v_x)^2 + s^3 \Theta^3 \frac{(x - x_0)^2}{a} v^2 \right) e^{2s\varphi} dx dt$$
$$\leq C \left(\int_0^T \int_0^1 h^2 e^{2s\varphi} dx dt + s c_1 \int_0^T \left[a\Theta e^{2s\varphi}(x - x_0)(v_x)^2 dt \right]_{x=0}^{x=1} \right).$$

For an analogous result in the Neumann case we refer to [20].

With suitable assumptions, from Carleman estimates and Theorem 1.3, one can prove observability inequalities and hence a related null controllability result. As promised, we refer to the original papers for the observability inequalities and the controllability result in the general case; here we quote a model result from [26].

Theorem 2.6 *The problem*

$$\begin{cases} u_t - (|x|^\alpha u_x)_x = f(t, x)\chi_\omega, & (t, x) \in (0, T) \times (-1, 1), \\ u(t, -1) = u(t, 1) = 0, & t \in (0, T), \\ u(0, x) = u_0(x), & x \in (-1, 1), \end{cases}$$

is null controllable

- *if $\alpha \in (0, 1)$ and $\omega = (a, b)$, or*
- *if $\alpha \in [1, 2)$ and $\omega = (a_1, b_1) \cup (a_2, b_2)$, $b_1 < 0 < a_2$.*

2.2.2   The Non Divergence Case

In the non divergence case we will consider the same spaces of Sect. 2.1.2 paying attention to the fact that a degenerates in an interior point x_0. For this reason, as before, we add x_0 in the notations. Again Theorem 2.3 holds requiring only that a is (WD) or (SD). On the contrary, in order to prove a Carleman estimate, we assume:

Hypothesis 2.5 The function a is weakly or strongly degenerate. Moreover,

$$\frac{(x - x_0)a'(x)}{a(x)} \in W^{1,\infty}(0, 1),$$

and, if $K_a \geq \dfrac{1}{2}$, then (2.28) holds. $\qquad\square$

Now, consider the function γ given in (2.25), where in this case

$$\psi(x) := d_1 \left(\int_{x_0}^{x} \frac{y - x_0}{a(y)} dy - d_2 \right), \qquad (2.34)$$

with $d_2 > \max \left\{ \dfrac{(1 - x_0)^2}{(2 - K_a)a(1)}, \dfrac{x_0^2}{(2 - K_a)a(0)} \right\}$ and $d_1 > 0$.

Theorem 2.7 (Theorem 4.2 [78]) *Assume Hypothesis 2.5. Then there exist two positive constants C and s_0 such that every solution*

$$v \in H^1\left(0, T; H^1_{\frac{1}{a}, x_0}(0, 1)\right) \cap L^2\left(0, T; H^2_{\frac{1}{a}, x_0}(0, 1)\right)$$

of

$$\begin{cases} v_t + a(x)v_{xx} = h, & (t, x) \in Q_T, \\ v(t, 1) = v(t, 0) = 0, & t \in (0, T), \end{cases}$$

satisfies

$$\int_0^T \int_0^1 \left(s\Theta v_x^2 + s^3 \Theta^3 \left(\frac{x - x_0}{a} \right)^2 v^2 \right) e^{2s\gamma} dxdt$$

$$\leq C \left(\int_0^T \int_0^1 h^2 \frac{e^{2s\gamma}}{a} dxdt + sd_1 \int_0^T \left[\Theta e^{2s\gamma}(x - x_0)v_x^2 dt \right]_{x=0}^{x=1} \right) \qquad (2.35)$$

for all $s \geq s_0$.

We underline the fact that to prove the previous result, inequality (2.31) is not necessary since all integrals and integrations by parts are justified by the definition of $H^2_{\frac{1}{a}, x_0}(0, 1)$, which can be written in a more appealing way as

$$H^2_{\frac{1}{a}, x_0}(0, 1) := \left\{ u \in H^1_{\frac{1}{a}, x_0}(0, 1) \, \middle| \, u' \in H^1(0, 1) \text{ and } au'' \in L^2_{\frac{1}{a}}(0, 1) \right\}.$$

2.3 Proof of Theorem 2.2

Let us finally prove the Carleman estimate given in Theorem 2.2. The lines of the proof below can be adapted, obviously even with major technical difficulties, to prove the other Carleman estimates stated in this notes. However, due to their elaborateness, this one will be the only proof we shall give for this types of inequalities. For it, we will follow [1].

We define, for $s > 0$, the function

$$w(t, x) := e^{s\varphi(t,x)} v(t, x),$$

where v is the solution of (2.7). Then w satisfies

$$\begin{cases} (e^{-s\varphi}w)_t + \left(a(x)(e^{-s\varphi}w)_x\right)_x = h, & (t, x) \in Q_T, \\ w(t, 1) = 0, & t \in (0, T), \\ \begin{cases} w(t, 0) = 0, & \text{for (\textbf{WD})}, \\ \text{or} & t \in (0, T), \\ (aw_x)(t, 0) = s(\varphi_x aw)(t, 0), & \text{for (\textbf{SD})}, \end{cases} \\ w(T, x) = w(0, x) = 0, & x \in (0, 1). \end{cases} \tag{2.36}$$

Thanks to the definitions of φ and ψ, we have

$$(\varphi_x aw)(t, x) = \Theta(t)a(x)\psi'(x)w(t, x) = c_1\Theta(t)xw(t, x).$$

Since, for $t \in [0, T]$, the function $w(\cdot, t)$ is in $H_a^1(0, 1)$, we deduce, using [1, Proposition 2.4], that:

$$xw(t, x)_{|x=0} = 0 \text{ for } t \in [0, T].$$

Thus

$$(\varphi_x aw)(t, x)_{|x=0} = 0 \text{ for } t \in [0, T],$$

and the previous problem can be recast as follows. Set

$$Lv := v_t + (a(x)v_x)_x \quad \text{and} \quad L_s w = e^{s\varphi}L(e^{-s\varphi}w), \quad s > 0.$$

Then (2.36) becomes

$$\begin{cases} L_s w = e^{s\varphi}h, \\ w(t, 1) = 0, & t \in (0, T), \\ \begin{cases} w(t, 0) = 0, & \text{for (\textbf{WD})}, \\ \text{or} & t \in (0, T), \\ (aw_x)(t, 0) = 0, & \text{for (\textbf{SD})}, \end{cases} \\ w(T, x) = w(0, x) = 0, & x \in (0, 1). \end{cases}$$

Computing $L_s w$, one has

$$L_s w = L_s^+ w + L_s^- w,$$

where $L_s^+ = \dfrac{L_s w + L_s^* w}{2}$ and $L_s^- = \dfrac{L_s w - L_s^* w}{2}$ denote the (formal) selfadjoint and skewadjoint parts of L_s. In this case

$$L_s^+ w := (a(x)w_x)_x - s\varphi_t w + s^2 a(x)\varphi_x^2 w,$$

and

$$L_s^- w := w_t - 2sa(x)\varphi_x w_x - s(a(x)\varphi_x)_x w.$$

Moreover

$$\|L_s^+ w\|_{L^2(0,1)}^2 + \|L_s^- w\|_{L^2(0,1)}^2 + 2 < L_s^+ w, L_s^- w >_{L^2(0,1)} = \|f e^{s\varphi}\|_{L^2(0,1)}^2.$$

For w the next result holds.

Proposition 2.5 *Let $T > 0$ be given. Then there exist two positive constants C and s_0, such that all solutions w of (2.36) satisfy, for all $s \geq s_0$,*

$$s \int_0^T \int_0^1 \Theta(t)a(x)w_x^2 dx dt + s^3 \int_0^T \int_0^1 \Theta^3(t)\frac{x^2}{a(x)}w^2 dx dt$$

$$\leq C \left(\int_0^T \int_0^1 h^2 e^{2s\varphi(t,x)} dx dt + sa(1)c_1 \int_0^T \Theta(t)w_x(t, 1)^2 dt \right),$$

where c_1 is the constant introduced in (2.10).

The proof of Proposition 2.5 is based on the computation of the scalar product $< L_s^+ w, L_s^- w >_{L^2(0,1)}$.

Lemma 2.3 *The following identity holds*

$$\int_0^T \int_0^1 L_s^+ w L_s^- w dx dt = \frac{s}{2} \int_0^T \int_0^1 \varphi_{tt} w^2 dx dt + s \int_0^T \int_0^1 a(x)(a(x)\varphi_x)_{xx} ww_x dx dt$$

$$- 2s^2 \int_0^T \int_0^1 a(x)\varphi_x \varphi_{tx} w^2 dx dt + s \int_0^T \int_0^1 (2a^2\varphi_{xx} + a(x)a'\varphi_x)w_x^2 dx dt$$

$$+ s^3 \int_0^T \int_0^1 (2a(x)\varphi_{xx} + a'\varphi_x)a(x)\varphi_x^2 w^2 dx dt + \int_0^T [a(x)w_x w_t]_{x=0}^{x=1} dt$$

$$+ \int_0^T [-s\varphi_x(a(x)w_x)^2 + s^2 a(x)\varphi_t\varphi_x w^2 - s^3 a^2\varphi_x^3 w^2]_{x=0}^{x=1} dt$$

$$+ \int_0^T [-sa(x)(a(x)\varphi_x)_x ww_x]_{x=0}^{x=1} dt.$$

$$(2.37)$$

Proof We have

$$\int_0^T \int_0^1 L_s^+ w w_t dx dt = \int_0^T \int_0^1 \{(a(x)w_x)_x - s\varphi_t w + s^2 a(x)\varphi_x^2 w\}w_t dx dt$$

$$= \int_0^T [a(x)w_x w_t]_{x=0}^{x=1} dt - \int_0^T \frac{1}{2}\frac{d}{dt}\left(\int_0^1 a(x)w_x^2 dx \right) dt$$

$$- \frac{s}{2} \int_0^T dt \int_0^1 \varphi_t (w^2)_t dx + \frac{s^2}{2} \int_0^T \int_0^1 a(x) \varphi_x^2 (w^2)_t dx dt$$

$$= \int_0^T [a(x) w_x w_t]_{x=0}^{x=1} dt + \frac{s}{2} \int_0^T \int_0^1 \varphi_{tt} w^2 dx dt \tag{2.38}$$

$$- s^2 \int_0^T \int_0^1 a(x) \varphi_x \varphi_{xt} w^2 dx dt,$$

since

$$\frac{s}{2} \int_0^1 [\varphi_t w^2]_{t=0}^{t=T} dx = \frac{s^2}{2} \int_0^1 [a(x) \varphi_x^2 w^2]_{t=0}^{t=T} dx$$

$$= \int_0^T \frac{1}{2} \frac{d}{dt} \left(\int_0^1 a(x) w_x^2 dx \right) dt = 0,$$

because of the choice of φ (we recall that $w(t = 0, x) = w(t = T, x) = 0$).
  In addition, we have

$$\int_0^T \int_0^1 L_s^+ w(-2sa(x)\varphi_x w_x) dx dt = -2s \int_0^T \int_0^1 \varphi_x \left[\frac{(a(x) w_x)^2}{2} \right]_x dx dt$$

$$+ 2s^2 \int_0^T \int_0^1 a(x) \varphi_t \varphi_x \left(\frac{w^2}{2} \right)_x dx dt - 2s^3 \int_0^T \int_0^1 a^2 \varphi_x^3 w w_x dx dt$$

$$= \int_0^T [-s \varphi_x (a(x) w_x)^2 + s^2 a(x) \varphi_t \varphi_x w^2 - s^3 a^2 \varphi_x^3 w^2]_{x=0}^{x=1} dt$$

$$+ s \int_0^T \int_0^1 \varphi_{xx} (a(x) w_x)^2 dx dt - s^2 \int_0^T \int_0^1 (a(x) \varphi_t \varphi_x)_x w^2 dx dt$$

$$+ s^3 \int_0^T \int_0^1 \{(a(x) \varphi_x^2)_x a(x) \varphi_x + a(x) \varphi_x^2 (a(x) \varphi_x)_x\} w^2 dx dt. \tag{2.39}$$

Moreover

$$\int_0^T \int_0^1 L_s^+ w(-s(a(x)\varphi_x)_x w) dx dt = \int_0^T [-sa(x) w_x w(a(x)\varphi_x)_x]_{x=0}^{x=1} dt$$

$$+ s \int_0^T \int_0^1 a(x) w_x \{(a(x)\varphi_x)_{xx} w + (a(x)\varphi_x)_x w_x\} dx dt$$

$$+ s^2 \int_0^T \int_0^1 (a(x)\varphi_x)_x \varphi_t w^2 dx dt \tag{2.40}$$

$$- s^3 \int_0^T \int_0^1 a(x) \varphi_x^2 (a(x)\varphi_x)_x w^2 dx dt.$$

Adding (2.38)–(2.40), (2.37) follows immediately.   □

The crucial step is to prove now the following estimate.

Lemma 2.4 *There exists a positive constant s_0 such that for all $s \geq s_0$ the distributed terms of (2.37) satisfy the following estimate*

$$\frac{s}{2} \int_0^T \int_0^1 \varphi_{tt} w^2 dx dt + s \int_0^T \int_0^1 a(x)(a(x)\varphi_x)_{xx} w w_x dx dt$$

$$- 2s^2 \int_0^T \int_0^1 a(x)\varphi_x \varphi_{tx} w^2 dx dt + s \int_0^T \int_0^1 (2a^2 \varphi_{xx} + a(x)a'\varphi_x) w_x^2 dx dt$$

$$+ s^3 \int_0^T \int_0^1 (2a(x)\varphi_{xx} + a'\varphi_x) a(x)\varphi_x^2 w^2 dx dt$$

$$\geq \frac{C}{2} s \int_0^T \int_0^1 \Theta(t) a(x) w_x^2 dx dt + \frac{C^3}{2} s^3 \int_0^T \int_0^1 \Theta^3(t) \frac{x^2}{a} w^2 dx dt,$$

for a positive constant C.

Proof Using the definition of φ, the distributed terms of $\int_0^T \int_0^1 L_s^+ w L_s^- w dx dt$ take the form

$$\frac{s}{2} \int_0^T \int_0^1 \ddot{\Theta}(t) \psi(x) w^2 dx dt + s \int_0^T \int_0^1 a(x)(a(x)\psi'(x))'' \Theta w w_x dx dt$$

$$- 2s^2 \int_0^T \int_0^1 \Theta(t) \dot{\Theta}(t) a(x)(\psi'(x))^2 w^2 dx dt \qquad (2.41)$$

$$+ s \int_0^T \int_0^1 \Theta(t) a(2a\psi''(x) + a'\psi'(x)) w_x^2 dx dt$$

$$+ s^3 \int_0^T \int_0^1 \Theta^3(t) a(2a\psi''(x) + a'\psi'(x))(\psi'(x))^2 w^2 dx dt.$$

Because of the choice of $\psi(x)$, one has $2a\psi''(x) + a'\psi'(x) = c_1 \frac{2a - xa'}{a}$ and $(a(x)\psi'(x))'' = 0$. Thus (2.41) becomes

$$\frac{s}{2} \int_0^T \int_0^1 \ddot{\Theta} \psi w^2 dx dt - 2s^2 \int_0^T \int_0^1 \Theta \dot{\Theta}(t) a(\psi')^2 w^2 dx dt$$

$$+ sc_1 \int_0^T \int_0^1 \Theta(t)(2a(x) - xa') w_x^2 dx dt$$

$$+ s^3 c_1 \int_0^T \int_0^1 \Theta^3(t)(\psi')^2 (2a(x) - xa') w^2 dx dt.$$

By assumption one can estimate the previous terms in the following way

$$\frac{s}{2} \int_0^T \int_0^1 \ddot{\Theta} \psi w^2 dx dt - 2s^2 \int_0^T \int_0^1 \Theta \dot{\Theta}(t) a (\psi')^2 w^2 dx dt$$

$$+ sc_1 \int_0^T \int_0^1 \Theta(t)(2a(x) - xa') w_x^2 dx dt$$

$$+ s^3 c_1 \int_0^T \int_0^1 \Theta^3(t)(\psi')^2 (2a(x) - xa') w^2 dx dt$$

$$\geq -2s^2 \int_0^T \int_0^1 \Theta \dot{\Theta}(t) a (\psi')^2 w^2 dx dt + \frac{s}{2} \int_0^T \int_0^1 \ddot{\Theta} \psi w^2 dx dt$$

$$+ sC \int_0^T \int_0^1 \Theta(t) a w_x^2 dx dt + s^3 C^3 \int_0^T \int_0^1 \Theta^3(t) \frac{x^2}{a} w^2 dx dt,$$

where $C > 0$ is some fixed constant.

Setting $b(x) = \int_0^x \frac{t}{a(t)} dt$ and since $|\Theta \dot{\Theta}| \leq c\Theta^{9/4} \leq c\Theta^3$ and $|\ddot{\Theta}| \leq c\Theta^{3/2}$, where c is a positive constant (see Remark 2.2), we can conclude that

$$\left| -2s^2 \int_0^T \int_0^1 \Theta(t) \dot{\Theta}(t) a (\psi')^2 w^2 dx dt \right| \leq 2cs^2 \int_0^T \int_0^1 \Theta^3(t) a (\psi')^2 w^2 dx dt$$

$$= 2cs^2 \int_0^T \int_0^1 \Theta^3(t) c_1^2 \frac{x^2}{a} w^2 dx dt \leq \frac{C^3}{4} s^3 \int_0^T \int_0^1 \Theta^3(t) \frac{x^2}{a} w^2 dx dt,$$

and

$$\left| \frac{s}{2} \int_0^T \int_0^1 \ddot{\Theta} \psi(x) w^2 dx dt \right| \leq \frac{s}{2} c_1 c \left| \int_0^T \int_0^1 \Theta^{3/2} b(x) w^2 dx dt \right|$$

$$+ s \frac{c_1 c_2}{2} c \left| \int_0^T \int_0^1 \Theta^{3/2} w^2 dx dt \right|,$$

for s large enough. Now, by Lemma 1.1, the function $x \mapsto \frac{x^K}{a(x)}$ is increasing, hence $b(x) \leq \frac{x^2}{(2-K) a(x)}$ and

$$\frac{s}{2} c_1 c \int_0^T \int_0^1 \Theta^{3/2} b(x) w^2 dx dt \leq \frac{C^3}{8} s^3 \int_0^T \int_0^1 \Theta^3 \frac{x^2}{a(x)} w^2 dx dt,$$

for s large enough.

It remains to bound the term $\left| \int_0^T \int_0^1 \Theta^{3/2} w^2 dx dt \right|$. At this step, we need to distinguish two cases. Indeed this part of the proof is based on the Poincaré–Hardy type inequalities (see Proposition 1.1). For this reason, the case $K = 1$ is peculiar.

Case $K \neq 1$. When $K \neq 1$ we have

$$s\frac{c_1c_2}{2}c\left|\int_0^1 \Theta^{3/2}w^2dx\right| = \frac{c_1c_2}{2}c\left|\int_0^1 \left(\sqrt{s}\Theta^{1/2}\frac{a^{\frac{1}{2}}}{x}w\right)\left(\sqrt{s}\Theta\frac{x}{a^{\frac{1}{2}}}w\right)dx\right|$$

$$\leq \frac{c_1c_2}{2}c\left(s\int_0^1 \Theta\frac{a}{x^2}w^2dx\right)^{\frac{1}{2}}\left(s\int_0^1 \Theta^2x^2a^{-1}w^2dx\right)^{\frac{1}{2}}$$

$$\leq \frac{c_1c_2}{2}c\left(C's\int_0^1 \Theta aw_x^2dx\right)^{\frac{1}{2}}\left(s\int_0^1 \Theta^2x^2a^{-1}w^2dx\right)^{\frac{1}{2}}$$

$$\leq \frac{\epsilon C's}{2}\int_0^1 \Theta a(x)w_x^2dx + \frac{C's}{2\epsilon}\int_0^1 \Theta^3(t)x^2a^{-1}w^2dx,$$

for some $C' > 0$. Then, for ϵ small enough and s large enough, we have

$$s\frac{c_1c_2}{2}c\left|\int_0^1 \Theta^{3/2}w^2dx\right| \leq \frac{C}{2}s\int_0^1 \Theta a(x)w_x^2dx + \frac{C^3}{8}s^3\int_0^1 \Theta^3\frac{x^2}{a(x)}w^2dx,$$

for the case $K \neq 1$.

Case K=1. When $K = 1$ we have to observe that

$$\int_0^1 w^2dx = \int_0^1 \left(w^{3/2}a^{1/4}x^{-1/2}\right)\left(w^{1/2}x^{1/2}a^{-1/4}\right)dx$$

$$= \int_0^1 \left(w^2a^{1/3}x^{-2/3}\right)^{3/4}\left(w^2x^2a^{-1}\right)^{1/4}dx$$

$$\leq \left(\int_0^1 w^2a^{1/3}x^{-2/3}dx\right)^{3/4}\left(\int_0^1 w^2x^2a^{-1}\right)^{1/4}$$

$$= \left(\int_0^1 w^2\left(\frac{a}{x^2}\right)^{1/3}dx\right)^{3/4}\left(\int_0^1 w^2x^2a^{-1}\right)^{1/4}.$$

Now, consider the function $p(x) = (a(x)x^4)^{1/3}$. It is clear that $p(x) \leq \frac{1}{a(1)^{2/3}}a(x)$. Moreover using the assumption on a, one can prove that $\frac{p(x)}{x^q}$ is nondecreasing near 0 for $q = \frac{4+\theta}{3}$. Thus we can apply the Hardy–Poincaré inequality given in Proposition 1.1, obtaining

$$\int_0^1 \left(\frac{a(x)}{x^2}\right)^{\frac{1}{3}}w^2(t,x)dx = \int_0^1 \frac{p(x)}{x^2}w^2(t,x)dx \leq C'\int_0^1 p(x)w_x^2(t,x)dx$$

$$\leq C'\int_0^1 a(x)w_x^2(t,x)dx,$$

for some positive constant C'. Thus

$$s \frac{c_1 c_2}{2} c \left| \int_0^1 \Theta^{3/2} w^2 dx \right|$$

$$\leq s \frac{c_1 c_2}{2} c \left| \Theta^{3/2} \left(C' \int_0^1 a(x) w_x^2(t, x) dx \right)^{3/4} \left(\int_0^1 w^2 x^2 a^{-1} dx \right)^{1/4} \right|$$

$$\leq s \frac{c_1 c_2}{2} C' \left(\int_0^1 \Theta a(x) w_x^2(t, x) dx \right)^{3/4} \left(\int_0^1 \Theta^3 x^2 a^{-1} w^2 dx \right)^{\frac{1}{4}}$$

$$\leq \frac{3 \epsilon C' s}{4} \int_0^1 \Theta a(x) w_x^2 dx + \frac{C' s}{4 \epsilon^3} \int_0^1 \Theta^3(t) x^2 a^{-1} w^2 dx,$$

for some $C' > 0$. Then, for ϵ small enough and s large enough, we have

$$s \frac{c_1 c_2}{2} c \left| \int_0^1 \Theta^{3/2} w^2 dx \right| \leq \frac{C}{2} s \int_0^1 \Theta a(x) w_x^2 dx + \frac{C^3}{8} s^3 \int_0^1 \Theta^3 \frac{x^2}{a(x)} w^2 dx,$$

for the case $K = 1$.

Using the above estimates for $K \neq 1$ and $K = 1$, we finally obtain

$$\left| \frac{s}{2} \int_0^T \int_0^1 \ddot{\Theta} \psi(x) w^2 dx dt \right| \leq \frac{C}{2} s \int_0^T \int_0^1 \Theta a(x) w_x^2 dx dt$$

$$+ \frac{C^3}{4} s^3 \int_0^T \int_0^1 \Theta^3 \frac{x^2}{a(x)} w^2 dx dt.$$

Summing up, we obtain

$$\frac{s}{2} \int_0^T \int_0^1 \varphi_{tt} w^2 dx dt + s \int_0^T \int_0^1 a(x)(a(x)\varphi_x)_{xx} w w_x dx dt$$

$$- 2s^2 \int_0^T \int_0^1 a(x)\varphi_x \varphi_{tx} w^2 dx dt + s \int_0^T \int_0^1 (2a^2 \varphi_{xx} + a(x)a'\varphi_x) w_x^2 dx dt$$

$$+ s^3 \int_0^T \int_0^1 (2a(x)\varphi_{xx} + a'\varphi_x)a(x)\varphi_x^2 w^2 dx dt$$

$$\geq \frac{C}{2} s \int_0^T \int_0^1 \Theta(t) a(x) w_x^2 dx dt + \frac{C^3}{2} s^3 \int_0^T \int_0^1 \Theta^3(t) \frac{x^2}{a} w^2 dx dt.$$

Lemma 2.5 *The boundary terms in (2.37) become*
Case (WD).

$$s \int_0^T [\Theta a^2 \psi' w_x^2]_{x=0} dt - sa(1)c_1 \int_0^T \Theta(t) w_x^2(t, 1) dt,$$

and

Case (SD).

$$-sa(1)c_1 \int_0^T \Theta(t)w_x^2(t, 1)dt,$$

where c_1 is the constant introduced in (2.10).

Proof Taking into account the fact that $w(t, 1) = 0$, one has

$$\text{(B.T.)} = -\int_0^T \left[aw_x w_t - s\Theta\psi'(aw_x)^2 + s^2\dot\Theta\Theta a\psi\psi'w^2 - s^3a^2\Theta^3(\psi')^3w^2\right]_{x=0} dt$$

$$+ \int_0^T \left[s\Theta(a\psi')'waw_x\right]_{x=0} dt - sa(1)c_1 \int_0^T \Theta(t)w_x^2(t, 1)dt.$$

In the **(WD)** case, we use the boundary condition $w(t, 0) = 0$ to obtain

$$\text{(B.T.)} = s\int_0^T \left[\Theta a^2\psi'w_x^2\right]_{x=0} dt - sa(1)c_1 \int_0^T \Theta(t)w_x^2(t, 1)dt.$$

On the other hand, in the **(SD)** case, we use the relation $(aw_x)(t, 0) = s\Theta(\psi'aw)(t, 0)$ to conclude that

$$\text{(B.T.)} = \int_0^T \left[-s\Theta a\psi'\left(\frac{w^2}{2}\right)_t - s^2\dot\Theta\Theta a\psi\psi'w^2 + 2s^3a^2\Theta^3(\psi')^3w^2\right.$$

$$\left.+ \Theta^2s^2w^2a\psi'(a\psi')'\right]_{x=0} dt - sa(1)c_1 \int_0^T \Theta(t)w_x^2(t, 1)dt$$

$$= \int_0^T \left[\frac{s}{2}\dot\Theta a\psi'w^2 - s^2\dot\Theta\Theta a\psi\psi'w^2 + 2s^3a^2\Theta^3(\psi')^3w^2\right.$$

$$\left.+ \Theta^2s^2w^2a\psi'(a\psi')'\right]_{x=0} dt - sa(1)c_1 \int_0^T \Theta(t)w_x^2(t, 1)dt.$$

Now, observe that

$$a\psi' = c_1x, \quad x \in (0, 1],$$

$$-a\psi\psi' \sim -c_1x\psi(0), \quad \text{as } x \to 0,$$

$$a^2(\psi')^3 = c_1^3\frac{x^3}{a(x)}, \quad x \in (0, 1],$$

$$a\psi'(a\psi')' = x, \quad x \in (0, 1].$$

Then

$$\text{(B.T.)} = \int_0^T \left[\left(c_1 \frac{s}{2} \dot{\Theta} - c_1 s^2 \dot{\Theta} \Theta \psi(0) + 2c_1^3 s^3 \Theta^3 \frac{x^2}{a} + c_1^2 \Theta^2 s^2 \right) (xw^2) \right]_{x=0} dt$$
$$- sa(1)c_1 \int_0^T \Theta(t) w_x^2(t, 1) dt.$$

Applying [1, Proposition 2.4], we have that $xw^2(t, x) \to 0$, as $x \to 0$. Moreover, using the fact that $\dfrac{x^2}{a} \leq \dfrac{1}{a(1)}$ (see Lemma 1.1), we can conclude that in case **(SD)**

$$\text{(B.T.)} = -sa(1)c_1 \int_0^T \Theta(t) w_x^2(t, 1) dt.$$

From Lemmas 2.3–2.5, we deduce directly the existence of two positive constants C and s_0, such that all solutions w of (2.36) satisfy, for all $s \geq s_0$,

$$\int_0^T \int_0^1 L_s^+ w L_s^- w \, dx \, dt \geq Cs \int_0^T \int_0^1 \Theta(t) a(x) w_x^2 \, dx \, dt$$
$$+ Cs^3 \int_0^T \int_0^1 \Theta^3(t) \frac{x^2}{a} w^2 \, dx \, dt - sa(1)c_1 \int_0^T \Theta(t) w_x^2(t, 1) \, dt, \tag{2.42}$$

where C is a positive constant depending on K and c_1 is the constant introduced in (2.10). This proves Proposition 2.5.

Recalling the definition of w, we have $v = e^{-s\varphi} w$ and $v_x = -s\Theta \psi' e^{-s\varphi} w + e^{-s\varphi} w_x$. Thus Theorem 2.2 follows immediately.

Chapter 3
A Singular Nondegenerate Parabolic Equation

Abstract We consider non degenerate singular parabolic problems, giving some existence or non existence results, which depend on the value of the parameter of the singular term. Null controllability results are presented, as well.

Keywords Singular equation · Subcritical, critical and supercritical cases · Null controllability

In this section we analyze the well posedness and the null controllability for the following problem

$$\begin{cases} u_t - \Delta u - \dfrac{\lambda}{|x|^2} u = f \chi_\omega, & (t, x) \in Q, \\ u(t, x) = 0, & (t, x) \in (0, T) \times \partial\Omega, \\ u(0, x) = u_0(x), & x \in \Omega, \end{cases} \tag{3.1}$$

where $u_0 \in L^2(\Omega)$, $f \in L^2(Q)$ and $\Omega \subset \mathbb{R}^N$ is a bounded open set such that $\partial\Omega$ is regular.

3.1 Well Posedness

The starting point for the well posedness of (3.1) is the paper of Baras and Goldstein [5]. In particular, they proved that, under suitable sign conditions, the existence of the solution depends on the parameter λ. Indeed, assume that Ω is an open subset of $\mathbb{R}^N$ such that

$$B_1 := \{x \in \mathbb{R}^N : |x| < 1\} \subset \Omega, \quad N > 1,$$
$$\Omega = (0, R), R \geq 1, \quad N = 1.$$

where $|\cdot|$ is the usual norm in $\mathbb{R}^N$. Consider two functions $\mathcal{V}$ and f such that $0 \leq \mathcal{V} \in L^1_{\text{loc}}(\Omega)$ and $0 \leq f \in L^1(Q)$. Let $u_0 \geq 0$ be a nontrivial function in $L^1(\Omega)$

or a finite (positive) Radon measure on Ω with $u_0 \not\equiv 0$. Baras and Goldstein focused on the existence of a solution for the following problem

$$
\begin{cases}
u_t - \Delta u - \mathcal{V}u = f, & \text{in } \mathscr{D}'(Q), \\
u \geq 0, & \text{on } Q, \\
\mathcal{V}u \in L^1_{\text{loc}}(Q), \\
\text{esslim}_{t \to 0^+} \int_\Omega u(t)\phi dx = \int_\Omega u_0 \phi dx, & \text{for all } \phi \in \mathscr{D}(\Omega).
\end{cases}
\tag{3.2}
$$

Here $\mathscr{D}(\Omega) = C_0^\infty(\Omega)$ is endowed with the standard topology and $\mathscr{D}'$ is the space of distributions. In order to treat the well posedness of (3.2), the authors considered the approximate problem

$$
\begin{cases}
(u_n)_t - \Delta u_n - \mathcal{V}_n u_n = f_n, & \text{in } \mathscr{D}'(Q), \\
u_n(t, x) = 0, & (t, x) \in (0, T) \times \partial\Omega, \\
\lim_{t \to 0^+} \int_\Omega u_n(t)\phi dx = \int_\Omega u_0\phi dx, & \text{for all } \phi \in \mathscr{D}(\Omega),
\end{cases}
\tag{3.3}
$$

where $f_n := \min\{f, n\}$ and $\mathcal{V}_n \in L^\infty(\Omega)$ is such that $0 \leq \mathcal{V}_n \leq \mathcal{V}$ and $\mathcal{V}_n \uparrow \mathcal{V}$ a.e. in Ω. For example, in the case that $\mathcal{V}(x) = \dfrac{\lambda}{|x|^2}$, as $\mathcal{V}_n$ one can consider $\mathcal{V}_n(x) := \min\{\mathcal{V}(x), n\}$. Then, problem (3.3) has a unique bounded non negative solution u_n given by

$$
u_n(t) = T(t)u_0 + \int_0^t T(t - s)\mathcal{V}_n u_n(s)ds + \int_0^t T(t - s)f_n(s)ds,
$$

where $(T(t))_{t \geq 0}$ is the semigroup generated by Δ with Dirichlet boundary conditions. Moreover, the sequence $\{u_n\}_n$ is increasing. Hence there exists $\lim_{n \to \infty} u_n$ and a non negative solution of (3.2) depends only on such a limit. Indeed, setting

$$
\lambda_* := \frac{(N - 2)^2}{4},
\tag{3.4}
$$

one has that when $\lambda \leq \lambda_*$ the sequence u_n converges monotonically to a solution u of the original problem, while for $\lambda > \lambda_*$ the sequence tends to infinity for all $(t, x) \in Q$. Hence, in this case we have *instantaneous* and *complete blow-up*. Summing up:

Proposition 3.1 (Proposition 2.1 [5]) *Let u_n solve (3.3).*

1. *Suppose that there exists $(t_0, x_0) \in Q$ such that $\lim_{n \to \infty} u_n(t_0, x_0) < \infty$. Then (3.2) has a non negative solution u on $(0, T_0) \times \Omega$ for all $T_0 \in (0, t_0)$ and*

$$
u(t, x) = \lim_{n \to \infty} u_n(t, x) \quad a.e. \text{ in } (0, T_0) \times \Omega.
$$

2. *If (3.2) has a non negative solution in Q, then*

$$\lim_{n \to \infty} u_n(t, x) < \infty \quad a.e. \; in \; (0, T_0) \times \Omega.$$

The borderline case is the potential given by

$$\mathcal{V}_0(x) = \begin{cases} \dfrac{\lambda}{|x|^2}, & \text{if } x \in B_1, \\ 0, & \text{if } x \in \Omega \setminus B_1, \end{cases}$$

as described by the following

Theorem 3.1 (Theorem 2.2 [5])

1. *Let $0 \le \lambda \le \lambda_*$ and the potential $\mathcal{V} \in L^\infty(\Omega \setminus B_1)$ be such that $\mathcal{V} \ge 0$. If $\mathcal{V} \le \mathcal{V}_0$ in B_1, then (3.2) has a solution u if*

$$\int_\Omega \frac{u_0}{|x|^\alpha} dx < \infty, \quad \int_Q \frac{f(s, x)}{|x|^\alpha} dx ds < \infty. \tag{3.5}$$

If $\mathcal{V} \ge \mathcal{V}_0$ in B_1 and if (3.2) has a solution u, then

$$\int_{\Omega'} \frac{u_0}{|x|^\alpha} dx < \infty, \quad \int_0^{T-\epsilon} \int_{\Omega'} \frac{f(s, x)}{|x|^\alpha} dx ds < \infty$$

for each $\epsilon \in (0, T)$ and each $\Omega' \subset\subset \Omega$. If either $u_0 \not\equiv 0$ or $f \not\equiv 0$ in $(0, \epsilon) \times \Omega$ for each $\epsilon \in (0, T)$, then given $\Omega' \subset\subset \Omega$ there is a constant $C = C(\epsilon, \Omega') > 0$ such that

$$u(t, x) \ge \frac{C}{|x|^\alpha} \quad if \; (t, x) \in [\epsilon, T) \times \Omega'.$$

   Here α is the smallest root of $(N - 2 - \alpha)\alpha = \lambda$.
2. *If $\lambda > \lambda_*$, $\mathcal{V} \ge \mathcal{V}_0$ and either $u_0 \not\equiv 0$ or $f \not\equiv 0$, then (3.2) does not have a solution.*

Observe that if $\phi(x) = \dfrac{1}{|x|^\alpha}$, then $\mathcal{V}(x) = -\dfrac{\Delta\phi}{\phi} = \dfrac{\lambda}{|x|^2}$, where λ and α are such that $\lambda = (N - 2 - \alpha)\alpha$, and indeed Theorem 3.1 can cover potentials of the form $\mathcal{V} = -\dfrac{\Delta\phi}{\phi}$ where $\phi > 0$, $\Delta\phi \in L^1_{loc}(\Omega)$, see [5, Remark 7.1].

The case $f \equiv 0$ is even more enlightening. Consider the problem

$$\begin{cases} u_t - \Delta u - \dfrac{\lambda}{|x|^2} u = 0, & (t, x) \in Q, \\ u(t, x) = 0, & (t, x) \in (0, T) \times \partial\Omega, \\ u(0, x) = u_0(x) \ge 0, & x \in \Omega, \\ u \ge 0, & \text{on } Q, \end{cases} \tag{3.6}$$

where $0 \not\equiv u_0 \in L^1(\Omega)$. Let u_n the unique solution of (3.6) with $\mathcal{V}_n(x) :=$ $\min\{\mathcal{V}(x), n\}$ in place of $\mathcal{V}(x) := \dfrac{\lambda}{|x|^2}$. As a consequence of the discussion at [5, p. 122], we have the following

Corollary 3.1 *1. If $0 \le \lambda \le \lambda_*$, then $\lim_{n\to\infty} u_n(t, x) = u(t, x)$ exists and u is a solution of (3.6).*
2. If $\lambda > \lambda_*$, then $\lim_{n\to\infty} u_n(t, x) = \infty$ for all $(t, x) \in Q$.

Observe that when $N = 2$ we have $\lambda_* = 0$. Hence, there is no $\lambda > 0$ for which the problem with Dirichlet boundary conditions and non negative initial datum has global solutions.

As a particular case of the general result established by Cabré and Martel in [24], we obtain the existence result for (3.6) with a different proof, essentially based on the Hardy inequality (1.16). For this, we recall the framework in [24]. Set $\delta(x) =$ $\text{dist}(x, \partial\Omega)$ for $x \in \Omega$, $L^1_\delta(\Omega) = L^1(\Omega, \delta(x)dx)$ and

$$\lambda_1(\mathcal{V}; \Omega) := \inf_{0\ne\phi\in C_c^\infty(\Omega)} \frac{\displaystyle\int_\Omega |\nabla\phi|^2 dx - \int_\Omega \mathcal{V}(x)\phi^2 dx}{\displaystyle\int_\Omega \phi^2 dx}.$$

We take the following definition from [24].

Definition 3.1 A function $u \ge 0$ is said to be a weak solution of (3.6) if for each $T_0 < T$ we have that $u \in L^1((0, T_0) \times \Omega)$, $\mathcal{V}u\delta \in L^1((0, T_0) \times \Omega)$ and

$$\int_0^{T_0} \int_\Omega u\left(-\varphi_t - \Delta\varphi\right) dxdt - \int_\Omega u_0\varphi(0)dx = \int_0^{T_0} \int_\Omega \mathcal{V}u\varphi dxdt,$$

for all $\varphi \in C^2([0, T_0] \times \overline{\Omega})$ with $\varphi(T_0) \equiv 0$ on Ω and $\varphi = 0$ on $[0, T_0] \times \partial\Omega$. If $T = \infty$, the solution u is said to be global.

The result in [24] applied to (3.6) reads as follows.

Theorem 3.2 (Theorem 1 [24])

1. *Suppose that, for some $u_0 \in L^1_\delta(\Omega)$ and some constants C and M, there exists a global weak solution $u \ge 0$ of (3.6) such that $\|u(t)\delta\|_{L^1(\Omega)} \le Ce^{Mt}$ for all $t \ge 0$. Then $\lambda_1(\mathcal{V}; \Omega) > -\infty$.*
2. *Suppose that $\lambda_1(\mathcal{V}; \Omega) > -\infty$. Then, for each $u_0 \in L^2(\Omega)$ with $u_0 \ge 0$, there exists a global weak solution $u \in C([0, \infty); L^2(\Omega))$ of (3.6) such that $\|u(t)\|_{L^2(\Omega)} \le \|u_0\|_{L^2(\Omega)}e^{-\lambda_1(\mathcal{V};\Omega)t}$ for all $t \ge 0$.*

Theorem 3.3 (Theorem 2 [24]) *Suppose that $\lambda_1((1 - \epsilon)\mathcal{V}; \Omega) = -\infty$ for some constant $\epsilon > 0$. Then, for any $T > 0$ and any $u_0 \in L^1_\delta(\Omega)$ with $u_0 \ge 0$ and $u_0 \not\equiv 0$, there is no weak solution $u \ge 0$ of (3.6). Moreover, there is instantaneous and complete blow-up for (3.6), in the following sense: for every $n \ge 1$, set $\mathcal{V}_n(x) = \min\{\mathcal{V}(x), n\}$*

and $u_{0,n}(x) = \min\{u_0(x); n\}$. Let u_n be the unique global solution of (3.6) associated to V_n and $u_{0,n}$. Then, for all $0 < T_0 < T$,

$$\frac{u_n(t, x)}{\delta(x)} \to +\infty \quad uniformly\ in\ (T_0, T) \times \Omega,\ as\ n \to \infty.$$

Applying directly these result, we re-obtain statements (i) and (ii) of Corollary 3.1 in the case of $0 \le u_0 \in L^2(\Omega)$ or $0 \le u_0 \in L^1_\delta(\Omega)$ and $u_0 \not\equiv 0$, respectively.

What happens if we remove the condition $u_0 \ge 0$? This problem, together with other questions, such as the asymptotic behavior or the uniqueness, is treated in [118]. In particular, Vazquez and Zuazua considered (3.6) without the condition $u \ge 0$, i.e.

$$\begin{cases} u_t - \Delta u - V(x)u = 0, & (t, x) \in Q, \\ u(t, x) = 0, & (t, x) \in (0, T) \times \partial\Omega, \\ u(0, x) = u_0(x), & x \in \Omega, \end{cases} \tag{3.7}$$

where Ω is a bounded domain of $\mathbb{R}^N$, $N \ge 3$, containing 0 and V is a locally integrable function defined in Ω with a singularity at one point, say 0. For example, as prototype, one can consider again

$$V(x) = \frac{\lambda}{|x|^2}, \quad \lambda \ge 0.$$

In order to study the well posedness of the previous problem, in [118] three cases are considered:

1. *Sub critical case:* $\lambda < \lambda_*$;
2. *Critical case:* $\lambda = \lambda_*$;
3. *Super critical case:* $\lambda > \lambda_*$,

where λ_* is defined in (3.4). Observe that, according to the previous results, the last case corresponds to the non existence of a solution $u \ge 0$ due to instantaneous blow-up. Hence, only the first two cases are interesting.

Sub Critical Case

Assume that V is such that $-C \le V(x) \le \dfrac{\lambda}{|x|^2}$ with $\lambda < \lambda_*$. Thanks to Theorem 1.1.1 one has that

$$\left(\int_\Omega (|\nabla u|^2 - V(x)u^2)dx\right)^{\frac{1}{2}}$$

is equivalent to the standard norm of $H_0^1(\Omega)$. Thus the operator $L : H_0^1(\Omega) \to H^{-1}(\Omega)$, defined as $L = -\Delta - V(x)I$ is an isomorphism. Hence, using the compactness of $H_0^1(\Omega)$ in $L^2(\Omega)$ and the imbedding of $L^2(\Omega)$ in $H^{-1}(\Omega)$, the operator L defines an unbounded self-adjoint operator in $L^2(\Omega)$ with compact inverse. As a consequence, there exist orthonormal eigenvectors of L, say $\{e_n\}_{n \in \mathbb{N}}$, which are a basis for $L^2(\Omega)$, and there is a sequence of associated eigenvalues $\{\mu_n\}_{n\in\mathbb{N}}$ such that

$$0 < \mu_1 \le \mu_2 \le \dots \to +\infty,$$

and

$$\begin{cases} -\Delta e_n - V(x)e_n = \mu_n e_n, & \text{in } \Omega, \\ e_n = 0, & \text{on } \partial\Omega. \end{cases}$$

Using this basis, one can prove

Theorem 3.4 (Theorem 3.2 [118]) *The operator L generates an analytic semigroup of contractions in $L^2(\Omega)$. Moreover, for any $u_0 \in L^2(\Omega)$ there exists a unique weak solution of problem (3.7) with*

$$u \in C[0, \infty); L^2(\Omega)) \cap L^2(0, \infty; H_0^1(\Omega)), \ u_t \in L^2(0, \infty; H^{-1}(\Omega))$$

and

$$u(t, x) = \sum_{n=1}^{\infty} a_n e^{-\mu_n t} e_n(x).$$

Here a_n, $n \ge 1$, are the Fourier coefficients of the initial data

$$u_0 = \sum_{n=1}^{\infty} a_n e_n.$$

Critical Case

Assume that V satisfies $-C \le V(x) \le \dfrac{\lambda_*}{|x|^2}$ but $V(x) \le \dfrac{\lambda}{|x|^2}$ is false for every $\lambda < \lambda_*$. In this case the Hardy–Poincaré inequality given in Theorem 1.1.1 is not sufficient to prove the coercivity of the differential operator L in $H_0^1(\Omega)$, but the improved Hardy–Poincaré inequality given in Theorem 1.1.2 suggests that problem (3.7) is well posed in a suitable Hilber space.

For this reason, the authors considered the Hilbert space H given by the completion of $\mathscr{D}(\Omega)$, or $H_0^1(\Omega)$, with respect to the norm

$$\|u\|_H := \left(\int_\Omega (|\nabla u|^2 - V(x)u^2)dx \right)^{\frac{1}{2}}$$

associated to the bilinear form

$$a(u, v) = \int_\Omega (\nabla u \cdot \nabla v - V(x)uv)dx.$$

For H the following embeddings hold:

$$H \hookrightarrow W_0^{1,q}(\Omega), \quad H \hookrightarrow\hookrightarrow H_0^s(\Omega)$$

if $1 \leq q < 2$ and $0 \leq s < 1$. The symbol $\hookrightarrow\hookrightarrow$ means that the second embedding is also compact due to the fact that $W_0^{1,q}(\Omega)$ is compactly embedded in $H_0^s(\Omega)$ for suitable $q = q(s)$ close enough to 2. Moreover, since $H_0^s(\Omega)$ is also compactly embedded in $L^2(\Omega)$, we have a compact embedding

$$H \hookrightarrow\hookrightarrow L^2(\Omega) \hookrightarrow H',$$

where H' is the dual space of H. Observe that when $V(x) = \dfrac{\lambda_*}{|x|^2}$, then H is larger than $H_0^1(\Omega)$ and smaller than $\bigcap_{q<2} W^{1,q}(\Omega)$ (for more details see [118]). In this setting, the same conclusion of Theorem 3.4 holds (see also [117, Theorem 2.1]):

Theorem 3.5 (Theorem 4.1 [118]) *For any $u_0 \in L^2(\Omega)$ there exists a unique weak solution of problem* (3.7) *with*

$$u \in C([0, \infty); L^2(\Omega)) \cap L^2(0, \infty; H), \ u_t \in L^2(0, \infty; H')$$

and u is given as in Theorem 3.4.

Super Critical Case

Assume now that $V(x) = \dfrac{\lambda}{|x|^2}$ with $\lambda > \lambda_*$. This situation is quite particular, and we shall not treat it here in details. For the sake of completeness, we recall that Theorem 3.1 states that in this case positive solutions blow up instantaneously. However, in [117], following the approach in [118], assuming that Ω is a ball, the authors proved that (3.7) is still well posed for a suitable subspace

$$H_\lambda$$

of initial conditions that oscillate sufficiently fast on the unit sphere, so that the singularity is compensated by such a fast oscillation.

Remark 3.1 The well posedness results stated above still hold true (with the required limitations in the supercritical case) for the problem

$$\begin{cases} u_t - \Delta u - V(x)u = f\chi_\omega, & (t, x) \in Q, \\ u(t, x) = 0, & (t, x) \in (0, T) \times \partial\Omega, \\ u(0, x) = u_0(x), & x \in \Omega, \end{cases}$$

for all $f \in L^2(Q)$, and thus for (3.1), see [117, Theorem 2.1].

The *super critical case* is considered also in [53]. Indeed, Theorem 3.1 states that if $\lambda > \lambda_*$ and $f = 0$ there is complete instantaneous blow-up for solutions to (3.1), which makes impossible to define a reasonable solution. But *given the initial datum u_0, can we find a control f localized in ω such that there exists a solution u of* (3.1)? In other words, *is it possible or impossible to prevent blow-up phenomena by acting* *only on the subset ω?* To give an answer to this question, in [53] the author considered the following functional: for any $u_0 \in L^2(\Omega)$ define

$$J_{u_0}(u, f) = \frac{1}{2} \int_0^T \int_\Omega |u(t, x)|^2 dx dt + \frac{1}{2} \int_0^T \|f(t)\|_{H^{-1}(\Omega)}^2 dt \qquad (3.8)$$

on the set

$$C(u_0) := \Big\{ (u, f) \in L^2((0, T); H_0^1(\Omega)) \times L^2((0, T); H^{-1}(\Omega)) : \text{such that}$$

$$u \text{ satisfies } (3.1) \text{ and } f \text{ is such that}$$

$$\forall \, \theta \in \mathcal{D}(\Omega \setminus \overline{\omega}), \; \theta f = 0 \text{ in } L^2((0, T); H^{-1}(\Omega)) \Big\},$$

$$(3.9)$$

and gave the following definition:

Definition 3.2 System (3.1) can be stabilized if there exists a constant $C > 0$ such that for all $u_0 \in L^2(\Omega)$

$$\inf_{(u, f) \in C(u_0)} J_{u_0}(u, f) \le C \|u_0\|_{L^2(\Omega)}^2. \qquad (3.10)$$

This property strongly depends on the set ω where the stabilization is effective. When $0 \in \omega$, (3.10) holds. Indeed, denote by φ a smooth function that equals 1 in a neighborhood of 0 and 0 outside ω and consider the solution u of

$$\begin{cases} u_t - \Delta u - (1 - \varphi) \dfrac{\lambda}{|x|^2} u = 0, & (t, x) \in Q, \\ u(t, x) = 0, & (t, x) \in (0, T) \times \partial \Omega, \\ u(0, x)) = u_0(x), & x \in \Omega. \end{cases}$$

By standard estimates for *non singular* problems, we get $u \in L^2(0, T; H_0^1(\Omega))$ and $\|u\|_{L^2(0,T;H_0^1(\Omega))} \le C \|u_0\|_{L^2(\Omega)}$ for some positive constant C. Then, $f = \lambda \varphi \dfrac{u}{|x|^2} \in L^2(0, T; H^{-1}(\Omega))$ is an admissible stabilizer that satisfies the required property (3.10). Moreover, in this case also null controllability holds, as well. Indeed, by classical controllability results (for instance, see [87]), we know that there exists a control $F \in L^2((0, T), \omega)$ such that the solution u of

$$\begin{cases} u_t - \Delta u - (1 - \varphi) \dfrac{\lambda}{|x|^2} u = F \chi_\omega, & (t, x) \in Q, \\ u(t, x) = 0, & (t, x) \in (0, T) \times \partial \Omega, \\ u(0, x)) = u_0(x), & x \in \Omega, \end{cases}$$

satisfies $u(T, \cdot) = 0$ in Ω. Moreover, there exists $C > 0$ such that

$$\|F\|_{L^2((0,T)\times\omega)} \le C\|u_0\|_{L^2(\Omega)} \text{ and } \|u\|_{L^2((0,T);H_0^1(\Omega))} \le C\|u_0\|_{L^2(\Omega)}.$$

In this way, taking $f = F + \lambda\varphi\dfrac{u}{|x|^2} \in L^2(0, T; H^{-1}(\Omega))$, we have a localized control which gives $u(T, \cdot) = 0$ in Ω.

When $0 \notin \overline{\omega}$ the situation is more involved. In this case, for $\epsilon > 0$, we approximate (3.1) by the system

$$\begin{cases} u_t - \Delta u - \dfrac{\lambda}{|x|^2 + \epsilon^2}u = f, & (t, x) \in Q, \\ u(t, x) = 0, & (t, x) \in (0, T) \times \partial\Omega, \\ u(0, x) = u_0(x), & x \in \Omega. \end{cases} \tag{3.11}$$

This Cauchy problem is well posed. Thus, we can consider the functionals

$$J_{u_0}^\epsilon(u, f) = \frac{1}{2}\int_0^T \int_\Omega |u(t, x)|^2 dx dt + \frac{1}{2}\int_0^T \|f(t)\|_{H^{-1}(\Omega)}^2 dt \tag{3.12}$$

where u satisfies (3.11) and f is localized in ω in the sense that

$$\forall \theta \in \mathscr{D}(\Omega \setminus \overline{\omega}), \ \theta f = 0 \text{ in } L^2((0, T); H^{-1}(\Omega)). \tag{3.13}$$

The following result holds

Theorem 3.6 (Theorem 1.2 [53]) *Assume that $\lambda > \lambda_*$ and that $0 \notin \overline{\omega}$. There is no constant C such that for all $\epsilon > 0$, and for all $u_0 \in L^2(\Omega)$,*

$$\inf_{\substack{f \in L^2(0, T; H^{-1}(\Omega)), \\ f \text{ as in } (3.13)}} J_{u_0}^\epsilon(f) \le C\|u_0\|_{L^2(\Omega)}^2.$$

This result implies that the stabilization of (3.1) is impossible to attain through regularization processes when $\lambda > \lambda_*$ and $0 \notin \overline{\omega}$. Moreover we cannot prevent the system from blowing up.

The well posedness in the case $0 \in \partial\Omega$, $\Omega \subset \mathbb{R}^N$, $N \ge 1$, is considered in [41]. In this case $f \in L^2(Q)$ is supported in a non-empty open region $\omega \subset \Omega$. To discuss well posedness, the crucial role is played by a new critical value of λ. More precisely, when moving the singularity from the interior to the boundary, the critical Hardy constant jumps from $\lambda_* = \dfrac{(N-2)^2}{4}$ to the new critical value $\lambda_N := \dfrac{N^2}{4}$, see Propositions 1.9 and 1.10, which are now the basic Hardy–Poincaré inequalities to prove well posedness.

To do that, let us fix $\gamma \in [0, 2)$, define the set

$$\mathcal{L}^\gamma := \left\{ C \geq 0 \ \text{s.t.} \ \inf_{u \in H_0^1(\Omega)} \frac{\displaystyle\int_\Omega \left(|\nabla u|^2 - \lambda_N \frac{u^2}{|x|^2} + C u^2 \right) dx}{\displaystyle\int_\Omega \frac{u^2}{|x|^\gamma} dx} \geq 1 \right\},$$

which is non empty by inequality (1.27) given in Proposition 1.9 (indeed $|C| \in \mathcal{L}^\gamma$), and consider the number

$$C_0^\gamma = \inf_{C \in \mathcal{L}^\gamma} C.$$

Then, for any $\lambda \leq \lambda_N$, we introduce the Hardy functional

$$B_\lambda^\gamma(u) := \int_\Omega |\nabla u|^2 dx - \lambda \int_\Omega \frac{u^2}{|x|^2} dx + C_0^\gamma \int_\Omega u^2 dx,$$

which is positive for any test function thanks to (1.27) and to the definition of C_0^γ. Then, we define the Hilbert space H_λ^γ as the closure of $C_0^\infty(\Omega)$ with respect to the norm induced by $B_\lambda^\gamma(\cdot)$. One can prove that if $\lambda < \lambda_N$, then H_λ^γ coincides with $H_0^1(\Omega)$, otherwise (if $\lambda = \lambda_N$) $H_{\lambda_N}^\gamma$ is slightly larger than $H_0^1(\Omega)$. On the space H_λ^γ, define the operator

$$A_\lambda^\gamma := -\Delta - \frac{\lambda}{|x|^2} + C_0^\gamma I$$

with domain

$$D(A_\lambda^\gamma) := \{ u \in H_\lambda^\gamma \mid A_\lambda^\gamma u \in L^2(\Omega) \}$$

and norm

$$\|u\|_{D(A_\lambda^\gamma)} = \|u\|_{L^2(\Omega)} + \|A_\lambda^\gamma u\|_{L^2(\Omega)} \ \forall \, \lambda \leq \lambda_N.$$

Then, by standard semigroup theory one can show that for any $\lambda \leq \lambda_N$ the operator $(A_\lambda^\gamma, D(A_\lambda^\gamma))$ generates an analytic semigroup in the space $L^2(\Omega)$ for the problem

$$\begin{cases} u_t - \Delta u - \dfrac{\lambda}{|x|^2} u = f, & (t, x) \in Q, \\ u(t, x) = 0, & (t, x) \in (0, T) \times \partial\Omega, \\ u(0, x) = u_0(x), & x \in \Omega, \end{cases}$$

which is thus well posed, see also [118].

3.2   Null Controllability for (3.1)

In this section we will analyze the null controllability property for (3.1). From the previous results, it is clear that one can expect null controllability only when $\lambda \leq \lambda_*$, if no other assumptions are made.

In [117], the authors considered the case when $0 \in \Omega \subset \mathbb{R}^N$, $N \geq 3$, and the control set ω is such that

$$\omega' := \{x \in \mathbb{R}^N : r_1 < |x| < r_2\} \subset \omega, \qquad (3.14)$$

for some constants $0 \leq r_1 < r_2$. Without loss of generality we can assume in the following that $r_2 < 1$.

The main result proved in [117] is the following

Theorem 3.7 (Theorem 2.2 [117]) *Assume that ω satisfies (3.14). The following statements hold:*

1. *If $\lambda \leq \lambda_*$, then for all $u_0 \in L^2(\Omega)$ there exists $f \in L^2(Q)$ such that the solution of (3.1) satisfies (1.42).*
2. *If $\lambda > \lambda_*$ and Ω is a ball, then for all $u_0 \in H_\lambda$ there exists $f \in L^2(Q)$ such that the solution of (3.1) satisfies (1.42).*

Here, for $\lambda > \lambda_*$, the space H_λ is the same one where the problem is well posed, see the **Super critical case** in the previous section.

Moreover, the previous theorem also gives a partial result on null controllability for the second case.

In order to prove Theorem 3.7, Vancostenoble and Zuazua used the equivalence stated in Theorem 1.3 between null controllability and observability inequality for the adjoint problem (when $f \equiv 0$)

$$\begin{cases} v_t + \Delta v + \dfrac{\lambda}{|x|^2} v = h, & (t, x) \in Q, \\ v(t, x) = 0, & (t, x) \in (0, T) \times \partial\Omega, \\ v(T, x) = v_T(x), & x \in \Omega, \end{cases} \qquad (3.15)$$

where v_T is given in $L^2(\Omega)$. In particular, (1.45) is proved in [117, Theorem 2.3] as a consequence of the following inequality:

$$\int_{\frac{T}{4}}^{\frac{3T}{4}} \int_\Omega v^2(t, x)dxdt \leq C \int_0^T \int_\omega v^2(t, x)dxdt. \qquad (3.16)$$

In this way, Vancostenoble and Zuazua are reduced to prove (3.16). To this aim, they split the domain in two subdomains: one containing the singularity of the potential and the other one in which the potential is bounded and smooth, i.e. near the boundary $\partial\Omega$. In this last region the problem can be studied by standard arguments using classical Carleman estimates. Thus, the main task is to study the problem on the region containing the singularity $x = 0$. The main idea is to decompose the N-dimensional problem (3.1) on spherical harmonics. Then the proof of (3.16) in the region near the singularity is reduced to proving some uniform Carleman estimates for an infinite family of 1-dimensional singular parabolic equations. In particular, taking $Q_T = (0, T) \times (0, 1)$ and considering the problem

$$\begin{cases} v_t + v_{xx} + \dfrac{\lambda}{x^2} v + \dfrac{m}{x^\beta} v = h, & (t, x) \in Q_T, \\ v(t, 0) = v(t, 1) = 0, & t \in (0, T), \\ v(T, x) = v_T(x) \in L^2(0, 1), & x \in (0, 1), \end{cases} \tag{3.17}$$

with $h \in L^2(Q_T)$, the next Carleman estimate is proved:

Theorem 3.8 (Theorem 3.2 [117]) *Assume* $\lambda \le \dfrac{1}{4}$, $\beta \in [0, 2)$ *and* $m \in \mathbb{R}$. *For every* $\gamma < 2$, *consider the function* $\sigma : (0, T) \times [0, 1] \to \mathbb{R}_+$ *given by*

$$\sigma(t, x) := \Theta(t)\left(1 - \frac{x^2}{2}\right), \quad \text{where } \Theta(t) := \left(\frac{1}{t(T-t)}\right)^k, \quad k := 1 + \frac{2}{\gamma}.$$

Then there exists $s_0 > 0$ *such that, for all* $s \ge s_0$, *the following inequality holds*

$$s^3 \int_0^T \int_0^1 \Theta^3 x^2 v^2 e^{-2s\sigma} dx dt + 2s \left(\frac{1}{4} - \lambda\right) \int_0^T \int_0^1 \Theta \frac{v^2}{x^2} e^{-2s\sigma} dx dt$$
$$+ \frac{s}{2} \int_0^T \int_0^1 \Theta \frac{v^2}{x^\gamma} e^{-2s\sigma} dx dt \le \frac{1}{2} \int_0^T \int_0^1 h^2 e^{-2s\sigma} dx dt$$

for the solutions v *of* (3.17) *satisfying the condition*

$$v(t, x) = 0 \quad \text{for all } (t, x) \in (0, T) \times (1 - \eta, 1), \text{ for some } 0 < \eta < 1. \tag{3.18}$$

Remark 3.2 Observe that condition (3.18) means that if v satisfies (3.17) then vanishes in a neighborhood of $x = 1$ and it is automatically satisfied by v when we prove (3.16) in the region containing the singularity by using cut off functions. However, taking into account that the singular potential is smooth near $x = 1$, one could also prove Carleman estimates for all solutions of (3.17) without assuming (3.18); in this way the constants appearing in these estimates would strongly depend on λ and blow up when $\lambda \to -\infty$.

Thanks to Theorem 3.8, one can prove that (3.1) and (3.17) can be controlled to zero with a distributed control which surrounds the singularity, i.e. in the special case when ω contains an annulus centered in the singularity (see (3.14)). As we have seen, this assumption is necessary for the technique that is used in [117]. However, in [53] it is proved that a such hypothesis can be removed and the equation can be null controlled from any open subset ω of Ω. Indeed, assuming that Ω is a smooth bounded domain of $\mathbb{R}^N$, $N \geq 3$, such that $0 \in \Omega$, the following theorem is proved.

Theorem 3.9 (Theorem 1.1 [53]) *Let $\lambda \in \mathbb{R}$ be such that $\lambda \leq \lambda_*$. Given any non-empty open set $\omega \subset \Omega$, for any $T > 0$ and $u_0 \in L^2(\Omega)$, there exists a control $f \in L^2((0, T) \times \omega)$ such that (3.1) is null controllable.*

Again, to prove Theorem 3.9, the author used Theorem 1.3 via a suitable Carleman estimate that is showed without using a spherical harmonics decomposition, avoiding the use of the geometrical condition (3.14). In particular, without loss of generality, in [53] it is assumed that $0 \notin \overline{\omega}$ (otherwise choose a smaller set ω), $\overline{B}_1 \subset \Omega$ and $\overline{B}_1 \cap \overline{\omega}$ is empty. As usual, the crucial problem to prove a Carleman estimate is a good choice for the weight function σ. In this case, the following weight is considered:

$$\sigma(t, x) = s\Theta(t)\left(e^{2\mu \sup \psi} - \frac{1}{2}|x|^2 - e^{\mu\psi(x)}\right),$$

where s and μ are sufficiently large constants and

$$\Theta(t) = \left(\frac{1}{t(T-t)}\right)^3.$$

Moreover, ψ is such that

$$\begin{cases} \psi(x) = \ln(|x|), & x \in B_1, \\ \psi(x) = 0, & x \in \partial\Omega, \\ \psi(x) > 0, & x \in \Omega \setminus \overline{B}_1 \end{cases}$$

and there exist an open set ω_0 and $\delta > 0$ so that $\overline{\omega}_0 \subset \omega$ and $|\nabla\psi(x)| \geq \delta$ for all $x \in \overline{\Omega} \setminus \omega_0$. The existence of such function is not straightforward, but it can be deduced following the construction in [87]. The choice of this function is due to the fact that in B_1, being ψ negative, we have that for μ large enough the function σ behaves like

$$s\Theta(t)\left(C - \frac{1}{2}|x|^2\right);$$

this corresponds to the weight considered in Theorem 3.8 when dealing with the observability around the singularity. On the other hand, outside the ball B_1, being ψ positive when μ is large, the weight is very close to the one used in [87]. Finally, observe that $\sigma(t, x) > 0$, for all $(t, x) \in (0, T) \times \Omega$ and $\lim_{t \to 0^+} \sigma(t, x) = \lim_{t \to T^-} \sigma(t, x) = +\infty$ for $x \in \Omega$. With this choice of the weight one has

Theorem 3.10 (Theorem 2.1 [53]) *There exist two positive constants C and μ_0 such that for all $\mu \geq \mu_0$, there exists $s_0 = s_0(\mu)$, such that for all $s \geq s_0$, any v solution of (3.15) with $h = 0$ satisfies*

$$s\mu^2 \int_0^T \int_{\tilde{O}} \Theta \phi e^{-2\sigma} |\nabla v|^2 dx dt + s \int_0^T \int_{\Omega} \Theta e^{-2\sigma} \frac{|v|^2}{|x|} dx dt$$

$$+ s^3 \int_0^T \int_{\Omega} \Theta^3 e^{-2\sigma} |x|^2 |v|^2 dx dt + s^3 \mu^4 \int_0^T \int_{\tilde{O}} \Theta^3 \phi^3 e^{-2\sigma} |v|^2 dx dt$$

$$\leq C \left(s\mu^2 \int_0^T \int_{\omega_0} \Theta \phi e^{-2\sigma} |\nabla v|^2 dx dt + s^3 \mu^4 \int_0^T \int_{\omega_0} \Theta^3 \phi^3 e^{-2\sigma} |v|^2 dx dt \right).$$

Here $\phi(x) = e^{\mu \psi(x)}$, $\mathcal{O} := \Omega \setminus (\overline{B}_1 \cup \overline{\omega}_0)$ and $\tilde{\mathcal{O}} := \Omega \setminus \overline{B}_1$.

The case $0 \in \partial \Omega$ is considered in [41], where the control region ω is allowed to be any open subset of Ω, no matter what the geometry of the domain is.

In particular, Cazacu proved the following result:

Theorem 3.11 (Theorem 1.3 [41]) *Let $\Omega \subset \mathbb{R}^N$, $N \geq 1$, be a domain such that $0 \in \partial \Omega$ and $\lambda \leq \lambda_*$. Given any non-empty open set $\omega \subset \Omega$, for any time $T > 0$ and any initial data $u_0 \in L^2(\Omega)$, there exists a control $f \in L^2((0,T) \times \omega)$ such that (3.1) is null controllable in the sense of Definition 1.5.*

This theorem is equivalent to Theorem 3.7.1 in the case $N = 1$. Indeed, we recall that using spherical harmonics decomposition, Vancostenoble and Zuazua reduce the problem in R^N, $N \geq 3$, to a one-dimensional problem in which the singularity arises at the origin and the control ω is distributed in an interval, say, $\Omega = (0, 1)$. Hence, the new case is for $N \geq 2$.

To prove the previous result, Cazacu used again the equivalence between null controllability and observability inequality given in Theorem 1.3, and, again, the mail tool is a suitable Carleman estimate. In this case, a good choice of σ is given by

$$\sigma(t, x) = \Theta(t) \left(C_\mu - |x|^2 \psi - \left(\frac{|x|}{r_0} \right)^\mu e^{\mu \psi(x)} \right).$$

Here μ is a positive large parameter, r_0 is chosen suitably small, C_μ is so large enough that $\sigma > 0$ and

$$\Theta(t) = \left(\frac{1}{t(T-t)} \right)^{1+\frac{2}{\gamma}},$$

$\gamma \in (1, 2)$ being the constant appearing in (1.27). As for the positive function ψ, it is constructed by fixing a nonempty subset $\omega_0 \subset\subset \omega$ in such a way that, finally, one has the following

Theorem 3.12 (Theorem 2.1 [41]) *There exist two positive constants C and μ_0 such that for all $\mu \geq \mu_0$, there exists $s_0 = s_0(\mu)$, such that for all $s \geq s_0$, any v solution of (3.15) satisfies*

$$s\mu^2 \int_0^T \int_{\mathcal{O}} \left(\frac{|x|}{r_0}\right)^{\mu} \Theta\phi e^{-2s\sigma}|\nabla v|^2 dxdt$$

$$+ s\int_0^T \int_{\Omega} \Theta e^{-2s\sigma}\left(|x|^{2-\gamma}|\nabla v|^2 + \frac{v^2}{|x|^{\gamma}}\right)dxdt$$

$$+ s^3\int_0^T \int_{\Omega} \Theta^3 e^{-2s\sigma}|x|^2 v^2 dxdt + s^3\mu^4\int_0^T \int_{\mathcal{O}} \Theta^3\phi^3 e^{-2s\sigma}\left(\frac{|x|}{r_0}\right)^{3\mu} v^2 dxdt$$

$$\leq Cs\mu^2\int_0^T \int_{\omega_0} \Theta\left(\frac{|x|}{r_0}\right)^{\mu} \phi e^{-2s\sigma}|\nabla v|^2 dxdt$$

$$+ Cs^3\mu^4\int_0^T \int_{\omega_0} \Theta^3\phi^3 e^{-2s\sigma}\left(\frac{|x|}{r_0}\right)^{3\mu} v^2 dxdt.$$

Here we have set

$$\phi := e^{\mu\psi(x)}, \qquad \mathcal{O} := \Omega \setminus (\overline{\omega}_0 \cup \overline{\tilde{\Omega}}_{r_0}),$$

$$\tilde{\Omega}_{r_0} := \Omega \cap B_{r_0} \text{ and } B_{r_0} := \{x \in \mathbb{R}^N : |x| < r_0\}.$$

Chapter 4
The Case of a Boundary Degenerate/Singular Parabolic Equation

Abstract We consider parabolic problems in divergence form with boundary degeneracy and power singularity, showing well posedness and null controllability.

Keywords Degenerate and singular equations · Well posedness · Null controllability

This chapter is devoted to study problems in which degenerate and singular operators coexist. In particular, we will treat the case in which the loss of uniform ellipticity occurs exactly in the same place where a singular potential appears. Though this situation might seem quite reductive, actually it is the most difficult one. Indeed, if the degeneracy and the singularity point do not coincide, they do not interact not even from a functional point of view, and so locally the problem is only degenerate or only singular, so known cases show up.

4.1 The Power Case

We start considering a model case studied in [115], given by the one dimensional degenerate/singular problem

$$\begin{cases} u_t - (x^\alpha u_x)_x - \dfrac{\lambda}{x^\beta} u = f, & (t, x) \in Q_T, \\ \begin{cases} u(t, 0) = 0 = u(t, 1), & \text{if } \alpha \in [0, 1), \\ (x^\alpha u_x)(t, 0) = 0 = u(t, 1), & \text{if } \alpha \geq 1, \end{cases} \\ u(0, x) = u_0(x), & x \in (0, 1), \end{cases} \tag{4.1}$$

where $\alpha \geq 0$ and $\lambda \in \mathbb{R} \setminus \{0\}$ satisfies suitable assumptions to be given below. Without loss of generality, we can assume that $\beta > 0$. Indeed, when $\beta \leq 0$, the potential is no more singular and the equation is purely degenerate, a case already treated in Chap. 2. Following [115], the boundary conditions vary according to the value of α: Dirichlet boundary conditions when $\alpha \in [0, 1)$ and some special Neumann-type boundary condition (see, for example, [25]) when $\alpha \geq 1$.

For this reason, in order to study well posedness and null controllability for problem (4.1) we distinguish two cases: $\alpha \in [0, 2) \setminus \{1\}$ and $\alpha = 1$. Observe that $\alpha \geq 2$ is not considered since in [37] it is proved that in this case null controllability is false (see also Chap. 2).

4.1.1   Well Posedness

When $\alpha = 0$, we have seen in the previous chapter that the critical exponent of the singular term $\dfrac{\lambda}{x^\beta}$ is $\beta = 2$. This is a consequence of the Hardy inequality (1.16) which holds for functions in $C_c^\infty(0, 1)$. In this degenerate/singular case, the critical exponent for the singular term decreases and takes into account the coexistence of both irregularities at the same point. Indeed, we will see that in this case the critical exponent becomes $\beta = 2 - \alpha$. This suggests to assume $\beta \leq 2 - \alpha$. Thus, we consider

$$\beta \in (0, 2 - \alpha] \text{ if } \alpha \in [0, 2) \setminus \{1\}$$

and λ *smaller than the related best Hardy constant*, since the critical exponent is included.

On the other hand, if $\alpha = 1$ we know from (1.17) that we cannot consider $\beta = 2 - \alpha = 1$, since the Hardy inequality makes no sense. However, if in (1.22) we choose $\gamma = \beta < 2 - \alpha = 1$, we have that, given $v \in C_c^\infty(0, 1)$, if $\sqrt{x}v_x \in L^2(0, 1)$, then $\dfrac{v}{x^{\frac{\beta}{2}}} \in L^2(0, 1)$ for all $\beta < 2 - \alpha = 1$. Hence, when $\alpha = 1$, we will study (4.1) considering

$$\beta \in (0, 1),$$

without condition on λ, since the critical case is excluded. Hence, following [115], we distinguish two cases:

1. *sub critical potentials:*

$$\begin{cases} \alpha \in [0, 2), \ \beta \in (0, 2 - \alpha), \quad \text{no condition on } \lambda; \\ \alpha \in [0, 2) \setminus \{1\}, \ \beta = 2 - \alpha, \quad \lambda < \lambda_*; \end{cases} \tag{4.2}$$

2. *critical potentials:*

$$\alpha \in [0, 2) \setminus \{1\}, \beta = 2 - \alpha, \lambda = \lambda_*. \tag{4.3}$$

Remark 4.1 Notice that, although in the second case of the sub critical case the critical growth is allowed, the associated energy functional is positive definite, since $\lambda < \lambda_*$. For this reason, we still call this situation sub critical, because the situation is simpler.

The Sub Critical Case

In order to study the well posedness of (4.1) in the *sub critical case* we consider the spaces given in Definition 1.3 with $a(x) = x^\alpha$. In particular:

Definition 4.1 If $\alpha \in [0, 1)$, define

$$\mathcal{H}_\alpha^1(0, 1) := \{u \in \mathcal{K}_\alpha^1(0, 1) : u(0) = u(1) = 0\},$$

and for $\alpha \in [1, 2)$ take

$$\mathcal{H}_\alpha^1(0, 1) := \{u \in \mathcal{K}_\alpha^1(0, 1) : u(1) = 0\},$$

where, in any case, $\mathcal{K}_\alpha^1(0, 1)$ is the Hilbert space

$$\mathcal{K}_\alpha^1(0, 1) := \{u \in L^2(0, 1) \cap H_{loc}^1(0, 1] : x^{\frac{\alpha}{2}} u' \in L^2(0, 1)\}$$

endowed with the scalar product

$$\langle u, v \rangle_{\mathcal{K}_\alpha^1} := \int_0^1 (uv + x^\alpha u' v') dx$$

for all $u, v \in \mathcal{K}_\alpha^1(0, 1)$.

Now, consider the operator $(A_\lambda, D(A_\lambda))$ given by

$$A_\lambda u := (x^\alpha u')' + \frac{\lambda}{x^\beta} u$$

for all $u \in D(A_\lambda)$, where

$$D(A_\lambda) := \{u \in \mathcal{H}_\alpha^1(0, 1) \cap H_{loc}^2(0, 1] : A_\lambda u \in L^2(0, 1)\}.$$

By definition of $\mathcal{H}_\alpha^1(0, 1)$ it is clear that if $\alpha \in [0, 1)$ and if $u \in D(A_\lambda)$ then u satisfies the Dirichlet boundary conditions $u(0) = u(1) = 0$. If $\alpha \in [1, 2)$ then u satisfies the Dirichlet boundary condition only at $x = 1$. However, Vancostenoble in [115, Proposition 1] proved that at $x = 0$ the function u satisfies the Neumann boundary condition $(x^\alpha u')(0) = 0$.

Using a semigroup approach, one can prove an existence result (see Theorem 4.1 below) based on the following inequality:

Proposition 4.1 (Proposition 2 [115]) *There exist $k \geq 0$ and $C > 0$ such that*

$$\int_0^1 \left(x^\alpha (u')^2 - \lambda \frac{u^2}{x^\beta} + ku^2 \right) dx \geq C \|u\|_{\mathcal{H}_\alpha^1}^2$$

for all $u \in \mathcal{H}_\alpha^1(0, 1)$.

Notice that, if $\alpha \neq 1$ and $\lambda < \lambda_*$, then $k = 0$.

Theorem 4.1 (Theorem 4.2 and Proposition 3 [115].) *Let $k \geq 0$ be the constant appearing in Proposition 4.1. Then $(A_\lambda - kI, D(A_\lambda))$ is a selfadjoint negative operator. Hence, if $f \equiv 0$ in (4.1), then for all $u_0 \in L^2(0, 1)$ problem (4.1) has a unique solution*

$$u \in C([0, T]; L^2(0, 1)) \cap C((0, T]; D(A_\lambda)) \cap C^1((0, T]; L^2(0, 1)).$$

Moreover, if $u_0 \in D(A_\lambda)$, then

$$u \in C([0, T]; D(A_\lambda)) \cap C^1([0, T]; L^2(0, 1)).$$

If $f \not\equiv 0$ and $f \in L^2(Q_T)$, then for all $u_0 \in L^2(0, 1)$ problem (4.1) has a unique solution

$$u \in C([0, T]; L^2(0, 1)).$$

Of course, estimate (1.40) and Theorem 1.2 apply.

The Critical Case

In the *critical case*, no analogue of Proposition 4.1 holds true. Hence, we have to modify the functional setting in a suitable way. In particular, in place of $\mathcal{K}_\alpha^1(0, 1)$, we consider the Hilber space $\mathcal{K}_\alpha^*(0, 1)$ given by

$$\mathcal{K}_\alpha^*(0, 1) := \left\{ u \in L^2(0, 1) \cap H^1_{\mathrm{loc}}(0, 1] : \int_0^1 \left(x^\alpha (u')^2 - \lambda_* \frac{u^2}{x^{2-\alpha}} \right) dx < +\infty \right\}$$

endowed with the scalar product

$$\langle u, v \rangle_{\mathcal{K}_\alpha^*} := \int_0^1 \left(uv + x^\alpha u'v' - \lambda_* \frac{uv}{x^{2-\alpha}} \right) dx.$$

As for the sub critical case, the trace at $x = 0$ makes sense if $\alpha < 1$ since one can prove that

$$\text{if } \alpha \in [0, 1), \text{ then } \mathcal{K}_\alpha^*(0, 1) \subset W^{1,1}(0, 1)$$

(see [115, Proposition 4]). Using $\mathcal{K}_\alpha^*(0, 1)$ we define the next spaces

Definition 4.2 1. For $\alpha \in [0, 1)$, define

$$\mathcal{H}_\alpha^*(0, 1) := \{ u \in \mathcal{K}_\alpha^*(0, 1) : u(0) = u(1) = 0 \}.$$

2. For $\alpha \in (1, 2)$ take

$$\mathcal{H}_\alpha^*(0, 1) := \{ u \in \mathcal{K}_\alpha^*(0, 1) : u(1) = 0 \}.$$

Again $\mathcal{H}_\alpha^*(0, 1)$ is the closure of $C_c^\infty(0, 1)$ with respect to the norm induced by $\langle \cdot, \cdot \rangle_{\mathcal{K}_\alpha^*}$ and (1.17), (1.22), (1.23) hold true in $\mathcal{H}_\alpha^*(0, 1)$. Therefore, thanks to (1.17), one can prove that

$$\|u\|_{\mathcal{H}_\alpha^*(0,1)} := \left(\int_0^1 \left(x^\alpha (u')^2 - \lambda_* \frac{u^2}{x^{2-\alpha}} \right) dx \right)^{\frac{1}{2}}$$

defines a norm on $\mathcal{H}_\alpha^*(0, 1)$ which is equivalent to $\| \cdot \|_{\mathcal{K}_\alpha^*}$.

In order to study the well posedness, we consider, as before, the operator $(A_{\lambda_*}, D(A_{\lambda_*}))$ with $\beta = 2 - \alpha$, but this time the domain is defined in the following way:

$$D(A_{\lambda_*}) := \{u \in \mathcal{H}_\alpha^*(0, 1) \cap H_{\mathrm{loc}}^2(0, 1] : A_{\lambda_*} u \in L^2(0, 1)\}$$

if $\alpha \in [0, 1)$ and

$$D(A_{\lambda_*}) := \{u \in \mathcal{H}_\alpha^*(0, 1) \cap H_{\mathrm{loc}}^2(0, 1] : A_{\lambda_*} u \in L^2(0, 1) \text{ and } (x^\alpha u')(0) = 0\}$$

if $\alpha \in (1, 2)$.

Let us notice that in [115, Proposition 5] it is proved that if $u \in D(A_{\lambda_*})$, then $x^\alpha u' \in W^{1,1}(0, 1)$. Then, in addition to the boundary conditions $u(0) = u(1) = 0$ in the first case, also the condition $(x^\alpha u')(0) = u(1) = 0$ in the second case makes sense, as well.

Now, we are ready for the well posedness result.

Theorem 4.2 (Proposition 6 [115]) *The operator $(A_{\lambda_*}, D(A_{\lambda_*}))$ is self-adjoint and negative. Hence, the analogue statements of Theorem* 4.1 *hold, as well.*

The proof of the previous result is an immediate consequence of the following equality, which replaces Proposition 4.1 in the previous sub critical situation:

$$\int_0^1 \left(x^\alpha (v')^2 - \lambda_* \frac{v^2}{x^\beta} \right) dx = \int_0^1 \left(x^\alpha (v')^2 - \lambda_* \frac{v^2}{x^{2-\alpha}} \right) dx = \|v\|_{\mathcal{H}_\alpha^*}^2,$$

$\forall v \in \mathcal{H}_\alpha^*(0, 1)$.

4.1.2 Null Controllability for (4.1)

As in the previous degenerate cases, also for this model we will deduce an observability inequality for the associated homogeneous adjoint problem to (4.1) via Carleman estimates which will be proved for the associated nonhomogeneous system

$$\begin{cases} v_t + A_\lambda v = h, & (t, x) \in Q_T, \\ v(t, 0) = 0 = v(t, 1), & \text{if } \alpha \in [0, 1), \\ (x^\alpha v_x)(t, 0) = 0 = v(t, 1), & \text{if } \alpha \geq 1, \\ v(T, x) = v_T(x), & x \in (0, 1). \end{cases} \tag{4.4}$$

To do that, define $\varphi(t, x) := \Theta(t)\psi(x)$, where

$$\Theta(t) := \left(\frac{1}{t(T-t)}\right)^k, \quad \text{with } k := 1 + \frac{2-\alpha}{\gamma}$$

and

$$\psi(x) := \frac{2 - x^{2-\alpha}}{(2-\alpha)^2}.$$

The next Carleman estimates hold:

Theorem 4.3 (Theorem 5.1 [115]) Let $\gamma \in (0, 2 - \alpha)$.

1. If $\alpha \in [0, 2)$, $\beta < 2 - \alpha$ and $\lambda \in \mathbb{R}$, then there exist $C > 0$ and $s_0 > 0$ such that, for all $s \geq s_0$, every solution v of (4.4) satisfies

$$\frac{s^3}{(2-\alpha)^3} \int_0^T \int_0^1 \Theta^3 x^{2-\alpha} v^2 e^{-2s\varphi} dx dt + s \int_0^T \int_0^1 \Theta \frac{v^2}{x^\gamma} e^{-2s\varphi} dx dt$$

$$+ s \int_0^T \int_0^1 \Theta \left(x^\alpha v_x^2 + (1-\alpha)^2 \frac{v^2}{x^{2-\alpha}}\right) e^{-2s\varphi} dx dt \tag{4.5}$$

$$\leq C \int_0^T \int_0^1 h^2 e^{-2s\varphi} dx dt + Cs \int_0^T \Theta[v^2 e^{-2s\varphi}]_{x=1} dt.$$

2. If $\alpha \in [0, 2) \setminus \{1\}$, $\beta = 2 - \alpha$ and $\lambda \leq \lambda_*$, then there exist $C > 0$ and $s_0 > 0$ such that, for all $s \geq s_0$, every solution v of (4.4) satisfies

$$\frac{s^3}{(2-\alpha)^3} \int_0^T \int_0^1 \Theta^3 x^{2-\alpha} v^2 e^{-2s\varphi} dx dt + s \int_0^T \int_0^1 \Theta \frac{v^2}{x^\gamma} e^{-2s\varphi} dx dt$$

$$+ s \int_0^T \int_0^1 \Theta \left(x^\alpha v_x^2 - \lambda \frac{v^2}{x^{2-\alpha}}\right) e^{-2s\varphi} dx dt \tag{4.6}$$

$$\leq C \int_0^T \int_0^1 h^2 e^{-2s\varphi} dx dt + Cs \int_0^T \Theta[v^2 e^{-2s\varphi}]_{x=1} dt.$$

Observe that by (1.17), inequality (4.6) implies that

$$s\left(1 - \frac{\lambda}{\lambda_*}\right) \int_0^T \int_0^1 \Theta x^\alpha v_x^2 e^{-2s\varphi} dxdt + (\lambda_* - \lambda) \int_0^T \int_0^1 \Theta \frac{v^2}{x^{2-\alpha}} e^{-2s\varphi} dxdt$$

$$\leq C \int_0^T \int_0^1 h^2 e^{-2s\varphi} dxdt + Cs \int_0^T \Theta[v^2 e^{-2s\varphi}]_{x=1} dt.$$

$$(4.7)$$

Let us notice that these two last inequalities generalize in some sense the Carleman estimate given in Theorem 3.8 in the case of the purely singular operator, that is $\alpha = 0$.

With the previous Carleman estimates on hand, one can prove an observability inequality like that in (1.45) with $C = C(2 - \alpha, \lambda, w)$, $\Omega = (0, 1)$ and w a nonempty subinterval of $(0, 1)$.

More precisely, let v be a solution of the homogeneous adjoint problem to (4.4), namely

$$\begin{cases} v_t + A_\lambda v = 0, & (t, x) \in Q_T, \\ v(t, 0) = 0 = v(t, 1), & \text{if } \alpha \in [0, 1), \\ (x^\alpha v_x)(t, 0) = 0 = v(t, 1), & \text{if } \alpha \geq 1, \\ v(T, x) = v_T(x), & x \in (0, 1). \end{cases} \quad (4.8)$$

Then, the following observability inequality holds.

Theorem 4.4 (Theorem 6.1 [115]) *Assume* (4.2) *or* (4.3). *Then there exists* $C = C(2 - \alpha, \lambda) > 0$ *such that for all* $v_T \in L^2(0, 1)$, *if* v *solves* (4.8), *then*

$$\int_0^1 v^2(0, x) dx \leq C \int_0^T \int_\omega v^2(t, x) dxdt. \quad (4.9)$$

In addition, if $\alpha \in [0, 2) \setminus \{1\}$, $\beta = 2 - \alpha$ *and* $\lambda \leq \lambda_*$, *then* $C(2 - \alpha, \lambda) = C(2 - \alpha)$.

By (4.9) and Theorem 1.3, we immediately get the following null controllability result.

Theorem 4.5 (Theorem 6.2 [115]) *Assume* (4.2) *or* (4.3) *and let* ω *be a nonempty subinterval of* $(0, 1)$. *Then the problem*

$$\begin{cases} u_t - (x^\alpha u_x)_x - \frac{\lambda}{x^\beta} u = f\chi_\omega, & (t, x) \in Q_T, \\ u(t, 0) = 0 = u(t, 1), & \text{if } \alpha \in [0, 1), \\ (x^\alpha u_x)(t, 0) = 0 = u(t, 1), & \text{if } \alpha \geq 1, \\ u(0, x) = u_0(x), & x \in (0, 1), \end{cases}$$

is null controllable.

4.2   A More General Function

4.2.1   Well Posedness

Now, we present a generalization of the previous results: in place of the pure power x^α in (4.1), consider a more general function a which satisfies Hypothesis 1.1. Hence, consider the following problem, studied in [68] and before in [69],

$$\begin{cases} u_t - (a(x)u_x)_x - \frac{\lambda}{x^\beta}u = f, & (t,x) \in Q_T, \\ u(t,0) = 0 = u(t,1), & \text{if } K_a \in [0,1), \\ (a(x)u_x)(t,0) = 0 = u(t,1), & \text{if } K_a \geq 1, \\ u(0,x) = u_0(x), & x \in (0,1), \end{cases} \tag{4.10}$$

where $u_0 \in L^2(0,1)$ and $f \in L^2(Q_T)$ are given functions.

In order to study problem (4.10), we make the following assumptions on K_a, β and λ: again, we assume (see [68])

$$K_a \in [0,2), \quad \beta \in (0, 2 - K_a), \quad \text{no condition on } \lambda \in \mathbb{R}, \tag{4.11}$$

or the more general one (see [69])

$$\begin{cases} K_a \in [0,2), \quad \beta \in (0, 2 - K_a), & \text{no condition on } \lambda \in \mathbb{R}; \\ K_a \in [0,2) \setminus \{1\}, \quad \beta = 2 - K_a, \quad \lambda < \lambda_*(a, K_a), \end{cases} \tag{4.12}$$

i.e. the case of the so-called *sub critical potentials* considered in the previous section. Here $\lambda_*(a, K_a)$ is the optimal constant appearing in the Hardy inequalities (1.20).

Concerning the well posedness, as before, we use a semigroup approach and, in analogy with the pure power case above, we consider the operator $(\mathcal{A}_\lambda, D(\mathcal{A}_\lambda))$ given by

$$\mathcal{A}_\lambda u := (a(x)u')' + \frac{\lambda}{x^\beta}u$$

for all $u \in D(\mathcal{A}_\lambda)$, where

$D(\mathcal{A}_\lambda) :=$
$$\begin{cases} \{u \in \mathcal{H}_a^1(0,1) \cap H^2_{\text{loc}}(0,1]; \ \mathcal{A}_\lambda u \in L^2(0,1)\}, & \text{if } K_a \in (0,1), \\ \{u \in \mathcal{H}_a^1(0,1) \cap H^2_{\text{loc}}(0,1]; \ \mathcal{A}_\lambda u \in L^2(0,1), (au')(0) = 0\}, & \text{if } K_a \in [1,2), \end{cases}$$

where, in this general case, the space $\mathcal{H}_a^1(0,1)$ is the one already introduced in Definition 1.3.

We notice some obvious facts. By definition of $\mathcal{H}_a^1(0,1)$ it is clear that if $K_a \in [0,1)$ and if $u \in D(\mathcal{A}_\lambda)$, then u satisfies the Dirichlet boundary conditions $u(0) =$

$u(1) = 0$. If $K_a \in [1, 2)$ then u satisfies the Dirichlet boundary condition only at $x = 1$, while at $x = 0$ it satisfies the Neumann boundary condition $(au')(0) = 0$. Indeed thanks to [68, Lemma 3.2] or to [69, Lemma 10], one has that if $K_a \in [1, 2)$ and (4.11) is satisfied, then for all $u \in \mathcal{K}_a^1(0, 1)$ such that $\mathcal{A}_\lambda u \in L^2(0, 1)$, $au' \in W^{1,1}(0, 1)$.

The following results hold.

Theorem 4.6 (Proposition 3 and Theorem 3.4 [68]) *Assume Hypothesis 1.1 and condition (4.11). Then there exists a constant $k \geq 0$ such that the operator $((\mathcal{A}_\lambda - kI), D(\mathcal{A}_\lambda))$ is a selfadjoint negative operator. Hence, the same conclusions of Theorem 4.1 still hold true.*

Theorem 4.7 (Proposition 12 and Theorem 14 [69]) *Assume Hypothesis 1.2 and condition (4.12). Then there exists a constant $k \geq 0$ such that the operator $((\mathcal{A}_\lambda - kI), D(\mathcal{A}_\lambda))$ is a selfadjoint negative operator. Hence, the same conclusions of Theorem 4.1 still hold true.*

In order to obtain the first part of the theorems above, and in particular the fact that the operator $(\mathcal{A}_\lambda - kI)$ is negative, it is essential to prove some generalizations of Proposition 4.1, whose proof is based on (1.25). For instance, when (1.19) holds, we have

Proposition 4.2 (Proposition 12 [69]) *Assume Hypothesis 1.2 and condition (4.12). Then there exist $k \geq 0$ and $C > 0$ such that*

$$\int_0^1 \left(a(u')^2 - \lambda \frac{u^2}{x^\beta} + ku^2 \right) dx \geq C \|u\|_{\mathcal{H}_a^1}^2$$

for all $u \in \mathcal{H}_a^1(0, 1)$.

4.2.2 Null Controllability

As usual in order to obtain null controllability for (4.10), it is sufficient to prove the observability inequality (1.45), with $\Omega = (0, 1)$ and ω a nonempty subinterval of $(0, 1)$, for the adjoint problem

$$\begin{cases} v_t + \mathcal{A}_\lambda v = 0, & (t, x) \in Q_T, \\ v(t, 0) = 0 = v(t, 1), & \text{if } K_a \in [0, 1), \\ (av_x)(t, 0) = 0 = v(t, 1), & \text{if } K_a \geq 1, \\ v(T, x) = v_T(x), & x \in (0, 1). \end{cases} \tag{4.13}$$

As usual, it is a consequence of Carleman estimates for the degenerate and singular problems

$$
\begin{cases}
v_t + \mathcal{A}_\lambda v = h, & (t,x) \in Q_T, \\
v(t,0) = 0 = v(t,1), & \text{if } K_a \in [0,1), \\
(av_x)(t,0) = 0 = v(t,1), & \text{if } K_a \geq 1, \\
v(T,x) = v_T(x), & x \in (0,1).
\end{cases}
\tag{4.14}
$$

To obtain these estimates, consider $\gamma \in (0, 2 - K_a)$, $\varphi(t,x) := \Theta(t)\psi(x)$, where

$$
\Theta(t) := \left(\frac{1}{t(T-t)} \right)^k, \quad \text{with } k := 1 + \frac{2}{\gamma}
$$

and

$$
\psi(x) := \frac{c_1}{2 - K_a} \left(\int_0^x \frac{y}{a(y)} \, dy - c_2 \right).
$$

Here c_1 is a suitable positive constant and $c_2 > 0$ is so large that $\psi(x) < 0$ for all $x \in [0,1]$. The next Carleman estimates hold.

Related to Theorem 4.6 we have

Theorem 4.8 (Theorem 4.2 [68]) *Assume Hypothesis* 1.1 *and* (4.11). *Then, there exist* $C > 0$ *and* $s_0 > 0$ *such that, for every* $h \in L^2(Q_T)$ *and* $v_T \in L^2(0,1)$, *if* v *solves* (4.14), *the following inequality holds for all* $s \geq s_0$:

$$
s^3 \int_0^T \int_0^1 \Theta^3 \frac{x^2}{a(x)} v^2 e^{2s\varphi} dx dt + s \int_0^T \int_0^1 \Theta \left(a(x) v_x^2 + \frac{v^2}{x^\beta} \right) e^{2s\varphi} dx dt
$$
$$
+ s \int_0^T \int_0^1 \Theta \frac{v^2}{x^\gamma} e^{2s\varphi} dx dt \leq C \int_0^T \int_0^1 f^2 e^{2s\varphi} dx dt + s \int_0^T \Theta [v_x^2 e^{2s\varphi}]_{x=1} dt.
$$

Associated to Theorem 4.7, we have

Theorem 4.9 (Theorem 17 [69]) *Assume Hypothesis* 1.2 *and condition* (4.12).

1. *If* $K_a \in (0,2)$, $\beta < 2 - K_a$ *and* $\lambda \in \mathbb{R}$, *then there exist* $C > 0$ *and* $s_0 > 0$ *such that, for all* $s \geq s_0$, *every solution* v *of* (4.14) *satisfies*

$$
\frac{s^3}{(2 - K_a)^2} \int_0^T \int_0^1 \Theta^3 \frac{x^2}{a(x)} v^2 e^{2s\varphi} dx dt + s \int_0^T \int_0^1 \Theta \frac{v^2}{x^\gamma} e^{2s\varphi} dx dt
$$
$$
+ s \int_0^T \int_0^1 \Theta \left(a(x) v_x^2 + \frac{a(1)(1 - K_a)^2}{4} \frac{v^2}{x^{2 - K_a}} \right) e^{2s\varphi} dx dt \tag{4.15}
$$
$$
\leq \frac{1}{2} \int_0^T \int_0^1 f^2 e^{2s\varphi} dx dt + \frac{3sa(1)}{2 - K_a} \int_0^T \Theta [v_x^2 e^{2s\varphi}]_{x=1} dt.
$$

2. *If* $K_a \in (0,2) \setminus \{1\}$, $\beta = 2 - K_a$ *and* $\lambda < \lambda_*$, *then there exist* $C > 0$ *and* $s_0 > 0$ *such that, for all* $s \geq s_0$, *every solution* v *of* (4.14) *satisfies*

$$\frac{s^3}{(2-K_a)^2}\int_0^T\int_0^1\Theta^3\frac{x^2}{a(x)}v^2e^{2s\varphi}dxdt + s\int_0^T\int_0^1\Theta\frac{v^2}{x^\gamma}e^{2s\varphi}dxdt$$

$$+\frac{sC}{2}\int_0^T\int_0^1\Theta\left(a(x)v_x^2 - \lambda\frac{v^2}{x^{2-K_a}}\right)e^{2s\varphi}dxdt \qquad (4.16)$$

$$\leq\frac{1}{2}\int_0^T\int_0^1 f^2e^{2s\varphi}dxdt + \frac{3sa(1)}{2-K_a}\int_0^T\Theta[v^2e^{2s\varphi}]_{x=1}dt,$$

where $C = \min\left\{1, \dfrac{\lambda_*(a, K_a) - \lambda}{|\lambda|}\right\}$.

Remark 4.2 Actually, both previous theorems were proved not only for problem (4.14), but for problems in which the equation contains a zero order term, more precisely $v_t + A_\lambda v - rv = h$ for some $r < 0$. This is not surprising, because it is well known that "lower order terms do not influence the Carleman estimates if the coefficients are bounded".[1]

With these Carleman estimates, the authors in [68, 69] proved related observability inequalities like that in (1.45). To do that, they need Hardy-type inequalities, like Lemma 1.3 and Proposition 1.8 in [68] and Lemma 1.2 and Proposition 1.7 in [69], and suitable Caccioppoli inequalities, like the following one:

Proposition 4.3 (Caccioppoli inequality, Lemma 28 [69]) *Let* $\omega' \subset\subset \omega$. *Then there exist* $C > 0$ *and* $s_0 > 0$ *such that every solution* v *of* (4.14) *satisfies*

$$\int_0^T\int_{\omega'}v_x^2e^{2s\varphi}dxdt \leq C\int_0^T\int_\omega v^2dxdt$$

for all $s \geq s_0$.

With an observability inequality at disposal, by Theorem 1.3, one has that null controllability holds, as well:

Theorem 4.10 (Theorem 4.1 [68]) *Assume Hypothesis* 1.1 *and* (4.11) *and let* ω *be a nonempty subset of* $(0, 1)$. *Then the system*

$$\begin{cases} u_t - (a(x)u_x)_x - \dfrac{\lambda}{x^\beta}u = f\chi_\omega, & (t, x) \in Q_T, \\[2mm] \begin{cases} u(t, 0) = 0 = u(t, 1), & \text{if } K_a \in [0, 1), \\ (a(x)u_x)(t, 0) = 0 = u(t, 1), & \text{if } K_a \geq 1, \end{cases} & \\[2mm] u(0, x) = u_0(x) \in L^2(0, 1), & x \in (0, 1), \end{cases} \qquad (4.17)$$

is null controllable, in the sense of Definition 1.5.

---

[1] See [10, p. 5].

Analogously, we have

Theorem 4.11 (Theorem 16 [69]) *Assume Hypothesis* 1.2 *and* (4.12) *and let* ω *be a nonempty subset of* (0, 1). *Then system* (4.17) *is null controllable, in the sense of Definition* 1.5.

4.3 A Special Case

We have seen above that there is a special case in which the Hardy inequality (1.17) is of no use, since it reduces to a trivial inequality, that is the case $\alpha = 1$. This means that the equation

$$u_t - (xu_x)_x - \lambda\frac{u}{x} = f \text{ in } Q_T$$

must be attacked in different ways. On the other hand, there are improved versions of the Hardy inequality, like (1.18) and (1.21). For instance, inequality (1.24) still holds if $v \in \mathcal{H}_\alpha^1(0, 1)$. Thus, if $v \in \mathcal{H}_\alpha^1(0, 1)$, then $\dfrac{v}{\sqrt{x}\ln x} \in L^2(0, 1)$. For this reason, one can consider problems with a logarithmic term $\lambda\dfrac{u}{x(\ln x)^2}$ in place of $\lambda\dfrac{u}{x}$, as done in [115]. So, let us consider

$$
\begin{cases}
u_t - (xu_x)_x - \dfrac{\lambda}{x(\ln x)^2}u = f, & (t, x) \in Q_T, \\
u(t, 0) = 0 = u(t, 1), & \text{if } \alpha \in [0, 1), \\
(x^\alpha u_x)(t, 0) = 0 = u(t, 1), & \text{if } \alpha \geq 1 \\
u(0, x) = u_0(x), & x \in (0, 1).
\end{cases}
\tag{4.18}
$$

Obviously, this problem is different from (4.1) since we have changed the potential. But, in order to study well posedness, we can proceed as before. In this case, thanks to (1.18), the critical threshold for λ is $\lambda_* = \dfrac{1}{4}$. This leads to consider $\lambda \leq \lambda_*$ and again we distinguish between two cases:

1. *sub critical potential:* $\lambda < \frac{1}{4}$,
2. *critical potential:* $\lambda = \frac{1}{4}$.

In this situation, the operator A_λ is replaced by B_λ, defined as

$$B_\lambda u := (xu_x)_x + \frac{\lambda}{x(\ln x)^2}u$$

for all $u \in D(B_\lambda)$. The functional setting, of course, is different in the sub critical and in the critical case: in the first case we have

$$D(B_\lambda) := \left\{ u \in \mathcal{H}_1^1(0, 1) \cap H^2_{\text{loc}}(0, 1] : \ B_\lambda u \in L^2(0, 1) \right\},$$

where $\mathcal{H}_1^1(0, 1)$ is the same space introduced in Definition 4.1, so that automatically $(xu')(0) = 0$. Concerning the critical case, we have

$$D(B_{\frac{1}{4}}) := \left\{ u \in \mathcal{H}_1^*(0, 1) \cap H^2_{\text{loc}}(0, 1] : \ B_{\frac{1}{4}} u \in L^2(0, 1) \text{ and } (xu')(0) = 0 \right\},$$

where, adapting Definition 4.2 to the case $\alpha = 1$, we now have

$$\mathcal{H}_1^*(0, 1) := \left\{ u \in L^2(0, 1) \cap H^1_{\text{loc}}(0, 1] : \right.$$

$$\left. \int_0^1 \left(x(u')^2 - \frac{1}{4} \frac{u^2}{x (\ln x)^2} \right) dx < +\infty \text{ and } u(1) = 0 \right\}$$

With these spaces, one can prove some results analogous to those established in Theorems 4.1 and 4.2. Thus, problem (4.18) is well posed in the setting established above.

Once well posedness is proved for (4.18), one can prove related Carleman estimates for the adjoint system. Thus, similarly to that one given in Theorem 4.3, we have the following result.

Theorem 4.12 (Theorem 6.4 [115]) *Consider* (4.18) *in the sub critical or in the critical case. Then, for all $\gamma \in (0, 1)$, there exists $s_0 = s_0(\gamma) > 0$ such that, for all $s \geq s_0$, every solution v of*

$$\begin{cases} v_t + B_\lambda v = hf, & (t, x) \in Q_T, \\ v(t, 0) = 0 = v(t, 1), & \text{if } \alpha \in [0, 1), \\ (x^\alpha v_x)(t, 0) = 0 = v(t, 1), & \text{if } \alpha \geq 1, \\ v(T, x) = v_T(x), & x \in (0, 1). \end{cases}$$

satisfies the inequality

$$s^3 \int_0^T \int_0^1 \Theta^3 x v^2 e^{-2s\varphi} dx dt + s \int_0^T \int_0^1 \Theta \left(x v_x^2 - \lambda \frac{v^2}{x (\ln x)^2} \right) e^{-2s\varphi} dx dt$$

$$+ s \int_0^T \int_0^1 \Theta \frac{v^2}{x^\gamma} e^{-2s\varphi} dx dt \leq \int_0^T \int_0^1 f^2 e^{-2s\varphi} dx dt + Cs \int_0^T \Theta [v^2 e^{-2s\varphi}]_{x=1} dt,$$

where $\varphi(t, x)$ is as in Theorem 4.3.

4.4   Further Developments

Degenerate, singular or simultaneously degenerate and singular equations have been considered in other different situations. Far from being exhaustive, we quote a list of other directions of research.

Hyperbolic operators: Control results for degenerate hyperbolic problems have been considered in [2] (boundary degeneracy), in [59] (degenerate or singular operator), in [12, 124] (boundary controller), in [123] (semilinear equation), in [125] (persistent regional null controllability[2]).

Parabolic systems: Carleman estimates and controllability results are proved in [11] (characterization of null controllability by the Kalman rank condition), in [119, 121] (cascade system), while in [93] stability conditions are established.

Nonlinear equations: Controllability results in [120] (equation in presence of a gradient term), in [50] (approximate boundary controllability for semilinear equations), in [28] (regional null controllability for semilinear equations).

---

[2] See Definition 2.2 for the parabolic case and [125] for the analogous definition for the degenerate wave equation.

Chapter 5
The Case of an Interior Degenerate/Singular Parabolic Equation

Abstract We consider parabolic problems in divergence and non divergence form with interior degeneracy and singularity given by general functions, showing well posedness and null controllability.

Keywords Degenerate and singular equations · Well posedness · Null controllability

We conclude these notes by treating situations which complement the ones studied in the previous chapter: we consider the evolution problem

$$\begin{cases} u_t - \mathcal{A}u - \dfrac{\lambda}{b(x)}u = f(t,x)\chi_\omega(x), & (t,x) \in Q_T, \\ u(t,0) = u(t,1) = 0, & t \in (0,T), \\ u(0,x) = u_0(x), & x \in (0,1), \end{cases} \tag{5.1}$$

where u_0, f belong to a suitable Hilbert space, $\lambda \in \mathbb{R}$, the control set ω satisfies Hypothesis 2.4 and $\mathcal{A}u = (a(x)u_x)_x$ or $\mathcal{A}u = a(x)u_{xx}$. In this case the functions a and b degenerate at the same *interior* point $x_0 \in (0,1)$. The fact that both a and b degenerate at the same point is the most complicated situation: indeed, if they degenerate at different points, the two phenomena can be faced separately, since there is no interaction between them. The prototypes we have in mind are

$$a(x) = |x - x_0|^{K_a} \text{ and } b(x) = |x - x_0|^{K_b} \text{ for some } K_a, K_b > 0.$$

The ways in which a and b degenerate at x_0 can be quite different, and for this reason we distinguish four different types of degeneracy. In particular, having in mind Definitions 1.1 and 1.2, we consider the following cases:

Definition 5.1 Doubly weakly degenerate case (WWD): if a, b are both **(WD)**.

Definition 5.2 Weakly-strongly degenerate case (WSD): if a is **(WD)** and b is **(SD)**.

© The Author(s), under exclusive license to Springer Nature Switzerland AG 2021
G. Fragnelli and D. Mugnai, *Control of Degenerate and Singular Parabolic Equations*,
SpringerBriefs in Mathematics, https://doi.org/10.1007/978-3-030-69349-7_5

Definition 5.3 Strongly-weakly degenerate case (SWD): if a is **(SD)** and b is **(WD)**.

Definition 5.4 Doubly strongly degenerate case (SSD): if a, b are both **(SD)**.

As in the case of a purely degenerate equation, the restriction $K_a, K_b < 2$ is related to the controllability issue. Indeed, it is clear from the proof of Theorem 5.4 below that such a condition is useless for the well posedness, for example, when $\lambda < 0$. On the other hand, concerning controllability, we will not consider the case $K_a, K_b \geq 2$, since if $a(x) = |x - x_0|^{K_a}$, $K_a \geq 2$ and $\lambda = 0$, by a standard change of variables (see [78]), problem (5.1) may be transformed in a non degenerate heat equation on an unbounded domain, while the control remains distributed in a bounded domain when $x_0 \notin \omega$. As we have seen in Chap. 2, this situation is now well understood, and the lack of null controllability was proved by Micu and Zuazua in [102].

We remark that in the previous chapters we have considered only a singular/degenerate operator with degeneracy or singularity appearing at the *boundary* of the domain. To the best of our knowledge, [20, 77, 78] are the first papers dealing with Carleman estimates (and, consequently, null controllability) for operators (in divergence and in non divergence form with Dirichlet or Neumann boundary conditions) with mere degeneracy at the *interior* of the space domain (for related systems of degenerate equations we refer to [19]).

We emphasize the fact that an interior degeneracy does not imply a simple adaptation of previous results and of the techniques used for boundary degeneracy. For instance, imposing homogeneous Dirichlet boundary conditions, in the latter case one knows a priori that any function in the reference functional space vanishes exactly at the degeneracy point. Now, since the degeneracy point is in the interior of the spatial domain, such information is not valid anymore, and we cannot take advantage of this fact. Moreover, one can show that, when the degeneracy point belongs to the control set and the coefficients are not too smooth, the controllability result cannot be obtained with standard methods (see [78, Chap. 7]).

5.1   The Divergence Form

For the well posedness of (5.1), we introduce the Hilbert space

$$\mathcal{H}^2_{a,b,x_0}(0, 1) := \left\{ u \in \mathcal{H}^1_{a,x_0}(0, 1) \; : \; au' \in H^1(0, 1) \text{ and } A_\lambda u \in L^2(0, 1) \right\},$$

where

$$A_\lambda u := \left(au'\right)' + \frac{\lambda}{b}u \quad \text{with} \quad D(A_\lambda) = \mathcal{H}^2_{a,b,x_0}(0, 1).$$

Observe that if $u \in D(\mathcal{A}_\lambda)$, then $\dfrac{u}{b}$ and $\dfrac{u}{\sqrt{b}} \in L^2(0, 1)$, so that $u \in \mathcal{H}_0$ and inequality (1.33) holds. We recall that $\mathcal{H}_0$ is defined in (1.35) and the other spaces are introduced in Sect. 1.2.

Moreover, we make the following assumptions on a, b and λ:

Hypothesis 5.1 1. The functions a and b are such that one of Definitions 5.1, 5.2 or 5.3 is satisfied with $K_a + K_b \leq 2$ and

$$\lambda \in \left(0, \frac{1}{C^*} \right), \tag{5.2}$$

or

2. The functions a and b are such that one among Definitions 5.1–5.3 or 5.4 holds with $\lambda < 0$. $\square$

Here C^* is the smallest constant for which (1.33) holds in $\mathcal{H}_0$. Observe that the assumption $\lambda \neq 0$ is not restrictive, since the case $\lambda = 0$ was already considered in Chap. 2.

Using Lemma 1.4 one can prove the next inequality.

Proposition 5.1 (Proposition 2.8 [79]) *Assume Hypothesis* 5.1. *Then there exists* $\Lambda \in (0, 1]$ *such that for all* $u \in \mathcal{H}_0$

$$\int_0^1 a(u')^2 dx - \lambda \int_0^1 \frac{u^2}{b} dx \geq \Lambda \int_0^1 a(u')^2 dx.$$

Thanks to Proposition 5.1 and the Green formula [84, Lemma 2.3], by using the semigroup theory, one can prove that (5.1) is well posed:

Theorem 5.1 (Theorem 2.22 [79]) *Assume Hypothesis* 5.1. *For every* $u_0 \in L^2(0, 1)$ *and* $f \in L^2(Q_T)$ *there exists a unique solution of problem* (5.1). *In particular, the operator* $\mathcal{A}_\lambda : D(\mathcal{A}_\lambda) \to L^2(0, 1)$ *is non positive and self-adjoint in* $L^2(0, 1)$. *Moreover, let* $u_0 \in D(\mathcal{A}_\lambda)$; *then*

$$f \in W^{1,1}(0, T; L^2(0, 1)) \Rightarrow u \in C^1(0, T; L^2(0, 1)) \cap C([0, T]; D(\mathcal{A}_\lambda)),$$
$$f \in L^2(Q_T) \Rightarrow u \in H^1(0, T; L^2(0, 1)).$$

Observe that, as we have seen in the previous chapters, the degenerate or non degenerate heat operator with an inverse-square singular potential gives rise to well posed Cauchy–Dirichlet problems if and only if λ is not larger than the best Hardy inequality (see [6, 24, 118]). For this reason, it is not strange that here we require an analogous condition for problem (5.1), by invoking Hypothesis 5.1.

In order to study null controllability for (5.1), we consider the non homogeneous adjoint problem

$$\begin{cases} v_t + (av_x)_x + \dfrac{\lambda}{b(x)}v = h(t, x), & (t, x) \in Q_T, \\ v(t, 0) = v(t, 1) = 0, & t \in (0, T), \\ v(T, x) = v_T(x). \end{cases} \tag{5.3}$$

On a and b we now need some additional assumptions:

Hypothesis 5.2 Hypotheses 2.3 and 5.1 hold. Moreover, if $\lambda < 0$ we require that

$$(x - x_0)b'(x) \ge 0 \text{ in } [0, 1].$$

To prove Carleman estimate, we consider the function φ given in (2.8), where ψ is as in (2.32). In this case, we have the following result.

Theorem 5.2 (Theorem 3.3 [79]) *Assume Hypothesis 5.2. Then, there exist two positive constants C and s_0, such that every solution v of (5.3) in*

$$\mathcal{V} := L^2\big(0, T; \mathcal{H}^2_{a,b,x_0}(0, 1)\big) \cap H^1\big(0, T; \mathcal{H}_0\big) \tag{5.4}$$

satisfies, for all $s \ge s_0$,

$$\int_0^T \int_0^1 \left(s\Theta a v_x^2 + s^3\Theta^3 \frac{(x - x_0)^2}{a} v^2 \right) e^{2s\varphi} dx dt$$

$$\le C \left(\int_0^T \int_0^1 h^2 e^{2s\varphi} dx dt + s \int_0^T \big[a\Theta e^{2s\varphi(t,x)}(x - x_0)v_x^2 dt \big]_{x=0}^{x=1} \right).$$

Remark 5.1 In [117] the authors proved a related Carleman inequality for the *non degenerate* singular 1-D problem

$$\begin{cases} v_t + v_{xx} + \dfrac{\mu}{x^2} + \dfrac{\lambda}{x^\beta}v = h & (t, x) \in Q_T, \\ v(t, 0) = v(t, 1) = 0 & t \in (0, T), \\ v(T, x) = v_T(x) & x \in (0, 1), \end{cases} \tag{5.5}$$

where $\beta \in [0, 2)$. When $\mu = 0$ and $x_0 = 0$, such an inequality reads simply as follows:

$$\int_0^T \int_0^1 \left(s^3\Theta^3 x^2 v^2 + \frac{s}{2}\Theta \frac{v^2}{x^2} + \frac{s}{2}\Theta \frac{v^2}{x^{2/3}} \right) e^{2s\psi} dx dt \le \frac{1}{2} \int_0^T \int_0^1 h^2 e^{2s\psi} dx dt,$$

where $\Psi(x) = \dfrac{x^2}{2} - 1 < 0$ in $[0, 1]$.

As a consequence of Theorem 5.2, with other technical assumptions (again satisfied by the prototypes) one can obtain the observability inequality for the homogeneous adjoint problem and immediately obtain a null controllability result, whose precise statement in the general situation can be found in [79, Theorem 4.5] and [83, Sect. 4]. Without entering into technical details, here we present the controllability result only in the prototype case, so that Hypothesis 5.2 is useless. So we have:

Theorem 5.3 *Assume Hypotheses 5.1 and 2.4. Then problem (5.1) with $a(x) = |x - x_0|^{K_a}$ and $b(x) = |x - x_0|^{K_b}$ is null controllable.*

We refer to [79] for the proof of (1.45) in this situation. Here we underline only that in order to prove it, a crucial role is played by (1.33). But, such an inequality doesn't hold when $K_a = K_b = 1$ (see [105]). Hence, with the technique used before, we are not able to prove the observability inequality (1.45). However, using cut-off functions and if the assumptions for the well posedness are satisfied, in [80] it is proved that null controllability for (5.1) with Dirichlet boundary conditions still holds in the **(SSD)** case, with general functions a and b. Hence, since the observability inequality is equivalent to null controllability by Theorem 1.3, we obtain that (1.45) still holds in the **(SSD)** case. Since the proof in this situation is different in nature from the previous analogous ones, we shall provide it for the following

Theorem 5.4 (Theorem 2.2 [80]) *Suppose that a and b are both **(SD)**, assume Hypothesis 2.4 and let $\lambda < 0$. Then, given $u_0 \in L^2(0, 1)$, there exists $f \in L^2(Q_T)$ such that the solution u of (5.1) satisfies*

$$u(T, x) = 0 \text{ for every } x \in [0, 1].$$

Moreover

$$\int_0^T \int_0^1 f^2 dx dt \le C \int_0^1 u_0^2(x) dx, \tag{5.6}$$

for some positive constant C. Hence, (1.45) holds.

Proof First, assume (2.29), that is $x_0 \in \omega \subset\subset (0, 1)$. Consider $0 < r' < r$ with $(x_0 - r, x_0 + r) \subset \omega$. Then, given an initial condition $u_0 \in L^2(0, 1)$, by classical controllability results in the non degenerate and non singular case, there exist two control functions $h_1 \in L^2((0, T) \times (0, x_0 - r'))$ and $h_2 \in L^2((0, T) \times (x_0 + r', 1))$, such that the corresponding solutions v_1 and v_2 of the parabolic problems

$$\begin{cases} v_t - (a(x)v_x)_x - \dfrac{\lambda}{b(x)} v = h_1 \chi_{\omega \cap (\zeta, x_0 - r)}(x), & (t, x) \in (0, T) \times (0, x_0 - r'), \\ v(t, 0) = v(t, x_0 - r') = 0, & t \in (0, T), \\ v(0, x) = u_0(x), & x \in (0, x_0 - r'), \end{cases} \tag{5.7}$$

and

$$\begin{cases} v_t - (a(x)v_x)_x - \dfrac{\lambda}{b(x)} v = h_2 \chi_{\omega \cap (x_0 + r, \xi)}(x), & (t, x) \in (0, T) \times (x_0 + r', 1), \\ v(t, x_0 + r') = v(t, 1) = 0, & t \in (0, T), \\ v(0, x) = u_0(x), & x \in (x_0 + r', 1), \end{cases}$$

respectively, satisfy $v_1(T, x) = 0$ for all $x \in (0, x_0 - r')$ and $v_2(T, x) = 0$ for all $x \in (x_0 + r', 1)$ with

$$\int_0^T \int_0^{x_0 - r'} h_1^2 dx dt \le C \int_0^T \int_0^{x_0 - r'} u_0^2 dx dt \tag{5.8}$$

and

$$\int_0^T \int_{x_0 + r'}^1 h_2^2 dx dt \le C \int_0^T \int_{x_0 + r'}^1 u_0^2 dx dt \tag{5.9}$$

for some constant C. Now, let u_3 be the solution of the problem

$$\begin{cases} v_t - (a(x)v_x)_x - \dfrac{\lambda}{b(x)} v = 0, & (t, x) \in Q_T, \\ v(t, 0) = v(t, 1) = 0, & t \in (0, T), \\ v(0, x) = u_0(x), & x \in (0, 1). \end{cases} \tag{5.10}$$

Denote by u_1 and u_2, f_1 and f_2 the trivial extensions of v_1 and v_2, h_1 and h_2 in $[x_0 - r', 1]$ and $[0, x_0 + r']$, respectively. Then take some cut-off functions $\phi_i \in C^\infty([0, 1])$, $i = 0, 1, 2$, with

$$\phi_1(x) := \begin{cases} 0, & x \in [x_0 - r', 1], \\ 1, & x \in [0, x_0 - r], \end{cases} \qquad \phi_2(x) := \begin{cases} 0, & x \in [0, x_0 + r'], \\ 1, & x \in [x_0 + r, 1], \end{cases}$$

and $\phi_0 = 1 - \phi_1 - \phi_2$. Finally, take

$$u(t, x) = \phi_1(x)u_1(t, x) + \phi_2(x)u_2(t, x) + \frac{T - t}{T} \phi_0(x)u_3(t, x). \tag{5.11}$$

Then, $u(T, x) = 0$ for all $x \in [0, 1]$ and u satisfies problem (5.1) in the domain Q_T with

$$f = \phi_1 f_1 \chi_{(\zeta, x_0 - r)} + \phi_2 f_2 \chi_{(x_0 + r, \xi)} - \frac{1}{T} \phi_0 u_3 - \phi_1' a u_{1,x} - \phi_2' a u_{2,x}$$
$$- \phi_0' \frac{T - t}{T} a u_{3,x} - \left(\phi_1' a u_1 + \phi_2' a u_2 + \phi_0' \frac{T - t}{T} a u_3 \right)_x.$$

Since a belongs to $W^{1,\infty}(0, 1)$, one has that $f \in L^2(Q_T)$, as required. Moreover, it is easy to see that the support of f is contained in ω.

Now, we prove (5.6). To this aim, consider the equation in (5.7) and multiply it by v_1. Then, integrating over $(0, x_0 - r')$, we have

$$\frac{1}{2}\frac{d}{dt}\|v_1(t)\|^2_{L^2(0,x_0-r')} + \|\sqrt{a}v_{1,x}(t)\|^2_{L^2(0,x_0-r')} - \lambda\left\|\frac{v_1}{\sqrt{b}}\right\|^2_{L^2(0,x_0-r')}$$

$$\leq \frac{1}{2}\|v_1(t)\|^2_{L^2(0,x_0-r')} + \frac{1}{2}\|h_1\|^2_{L^2(\omega\cap(\zeta,x_0-r))}.$$

Using the fact that $\lambda < 0$, we find

$$\frac{d}{dt}\|v_1(t)\|^2_{L^2(0,x_0-r')} \leq \frac{d}{dt}\|v_1(t)\|^2_{L^2(0,x_0-r')} + 2\|\sqrt{a}v_{1,x}(t)\|^2_{L^2(0,x_0-r')}$$

$$\leq \|v_1(t)\|^2_{L^2(0,x_0-r')} + \|h_1(t,\cdot)\|^2_{L^2(\omega\cap(\zeta,x_0-r))}.$$

Integrating the previous inequality, we get

$$\|v_1(t)\|^2_{L^2(0,x_0-r')} \leq e^T\left(\|u_0\|^2_{L^2(0,x_0-r')} + \int_0^t \|h_1(t,\cdot)\|^2_{L^2(\omega\cap(\zeta,x_0-r))}\right)$$

for all $t \in [0, T]$, and so

$$\|v_1\|^2_{L^2((0,x_0-r')\times[0,T])} \leq C\left(\|u_0\|^2_{L^2(Q_T)} + \|h_1\|^2_{L^2((0,x_0-r')\times[0,T])}\right). \tag{5.12}$$

Now, integrating over $(0, T)$ the inequality

$$\frac{d}{dt}\|v_1(t)\|^2_{L^2(0,x_0-r')} + 2\|\sqrt{a}v_{1,x}(t)\|^2_{L^2(0,x_0-r')}$$

$$\leq \|v_1(t)\|^2_{L^2(0,x_0-r')} + \|h_1(t,\cdot)\|^2_{L^2(\omega\cap(\zeta,x_0-r))},$$

by using (5.12), we immediately find

$$\|\sqrt{a}v_{1,x}\|^2_{L^2((0,x_0-r')\times[0,T])} \leq C\left(\|u_0\|^2_{L^2(Q_T)} + \|h_1\|^2_{L^2((0,x_0-r')\times[0,T])}\right) \tag{5.13}$$

for some $C > 0$.

Now, let us note that, since $a \in W^{1,\infty}(0, 1)$, then

$$\|(av_1)_x\|_{L^2((0,x_0-r')\times[0,T])}$$

$$\leq C\left(\|v_1\|_{L^2((0,x_0-r')\times[0,T])} + \|\sqrt{a}v_{1,x}\|_{L^2((0,x_0-r')\times[0,T])}\right).$$

By using (5.12) and (5.13) in the previous inequality, we get

$$\|(av_1)_x\|_{L^2((0,x_0-r')\times[0,T])} \leq C\left(\|u_0\|^2_{L^2(Q_T)} + \|h_1\|^2_{L^2((0,x_0-r')\times[0,T])}\right) \tag{5.14}$$

for some $C > 0$.

An estimate analogous to (5.14) holds for v_2 with h_2 replacing h_1, and for u_3 only in terms of u_0 in their own domains.

In conclusion, by (5.12)–(5.14), from the very definition of f and by (5.8) and (5.9), inequality (5.6) follows immediately.

Now, assume (2.30). Take $r > 0$ such that $\xi_1 < x_0 - r$ and $x_0 + r < \zeta_2$. As before, given an initial condition $u_0 \in L^2(0, 1)$, by classical controllability results in the non degenerate and non singular case, there exist two control functions $h_4 \in L^2((0, T) \times (0, x_0 - r))$ and $h_5 \in L^2((0, T) \times (x_0 + r, 1))$, such that the corresponding solutions v_4 and v_5 of the parabolic problems

$$
\begin{cases}
v_t - (a(x)v_x)_x - \dfrac{\lambda}{b(x)}v = h_1(t, x)\chi_{(\zeta_1, \xi_1)}(x), & (t, x) \in (0, T) \times (0, x_0 - r), \\
v(t, 0) = v(t, x_0 - r) = 0, & t \in (0, T), \\
v(0, x) = u_0(x), & x \in (0, x_0 - r),
\end{cases}
\tag{5.15}
$$

and

$$
\begin{cases}
v_t - (a(x)v_x)_x - \dfrac{\lambda}{b(x)}v = h_2(t, x)\chi_{(\zeta_2, \xi_2)}(x), & (t, x) \in (0, T) \times (x_0 + r, 1), \\
v(t, x_0 + r) = v(t, 1) = 0, & t \in (0, T), \\
v(0, x) = u_0(x), & x \in (x_0 + r, 1),
\end{cases}
\tag{5.16}
$$

respectively, satisfy $v_4(T, x) = 0$ for all $x \in (0, x_0 - r)$ and $v_5(T, x) = 0$ for all $x \in (x_0 + r, 1)$ with

$$
\int_0^T \int_0^{x_0 - r} h_4^2 \, dx \, dt \le C \int_0^T \int_0^{x_0 - r} u_0^2 \, dx \, dt
\tag{5.17}
$$

and

$$
\int_0^T \int_{x_0 + r}^1 h_5^2 \, dx \, dt \le C \int_0^T \int_{x_0 + r}^1 u_0^2 \, dx \, dt
\tag{5.18}
$$

for some constant $C > 0$. As before, let u_4 and f_4, u_5 and f_5 be the trivial extensions of v_4 and h_4, v_5 and h_5 in $[x_0 - r, 1]$ and $[0, x_0 + r]$, respectively.

Then, define cut-off functions $\varphi_i \in C^\infty([0, 1])$, $i = 0, 1, 2$, such that

$$
\varphi_1(x) := \begin{cases} 0, & x \in [\xi_1, 1], \\ 1, & x \in [0, \zeta_1], \end{cases}
\qquad
\varphi_2(x) := \begin{cases} 0, & x \in [0, \zeta_2], \\ 1, & x \in [\xi_2, 1], \end{cases}
$$

and $\varphi_0 = 1 - \varphi_1 - \varphi_2$. Finally, set

$$
u(t, x) = \varphi_1(x)u_4(t, x) + \varphi_2(x)u_5(t, x) + \frac{T - t}{T}\varphi_0(x)u_3(t, x),
\tag{5.19}
$$

where u_3 is the solution of (5.10).

As before, $u(T, x) = 0$ for all $x \in [0, 1]$ and u satisfies problem (5.1) in the domain Q with

$$
f = \varphi_1 f_4 \chi_{(\zeta_1, \xi_1)} + \varphi_2 f_5 \chi_{(\zeta_2, \xi_2)} - \frac{1}{T} \varphi_0 u_3 - \varphi_1' a u_{4,x} - \varphi_2' a u_{5,x}
$$
$$
- \varphi_0' \frac{T - t}{T} a u_{3,x} - \left(\varphi_1' a u_4 + \varphi_2' a u_5 + \varphi_0' \frac{T - t}{T} a u_3 \right)_x .
$$

Again $f \in L^2(Q_T)$, as required and the support of f is contained in ω. In order to conclude we have to prove (5.6) for the control function f, but such an estimate can be obtained as above, and the conclusion follows.   □

We underline that, in [80], Theorem 5.4 is proved also when a is **(WD)**. In this case, even if $a \in C[0, 1] \cap (C^1[0, 1] \setminus \{x_0\})$, the function f that we find above in general *is not* in $L^2(Q_T)$ without any additional assumption. Moreover, this technique works also in the **(SWD)** provided that the there exists a solution of (5.1), for example if $\lambda < 0$ or $\lambda > 0$ small enough and $K_a + K_b \leq 2$, thus we re-obtain the controllability result given in [79, Theorem 2.22].

The importance of Theorem 5.4 is clarified also by the following observation.

Remark 5.2 The null controllability result in Theorem 5.4 *cannot be obtained* by results already known in literature. Indeed, one may think to consider

$$
\begin{cases}
u_t - (a(x)u_x)_x - \dfrac{\lambda}{b(x)} u = f(t, x) \chi_{(\zeta_1, \xi_1)}(x), & (t, x) \in (0, T) \times (0, x_0), \\
u(t, 0) = u(t, x_0) = 0, & t \in (0, T), \\
u(0, x) = u_0(x)_{|[0, x_0)},
\end{cases}
$$
                                                                                                (5.20)

and

$$
\begin{cases}
u_t - (a(x)u_x)_x - \dfrac{\lambda}{b(x)} u = f(t, x) \chi_{(\zeta_2, \xi_2)}(x), & (t, x) \in (0, T) \times (x_0, 1), \\
u(t, x_0) = u(t, 1) = 0, & t \in (0, T), \\
u(0, x) = u_0(x)_{|(x_0, 1]},
\end{cases}
$$
                                                                                                (5.21)

and say that u is a solution of (5.1) if and only if the restrictions of u to $[0, x_0)$ and to $(x_0, 1]$, are solutions to (5.20) and (5.21), respectively. Thanks to [79, Lemma 2.11], which is a characterization of the space $\mathcal{H}_0(0, 1)$ defined in (1.35), if ω satisfies (2.30) and the initial datum is more regular, this can actually be done. Indeed, in this case we have two problems with degeneracy and singularity at the boundary. However, in this case the only available results are, for instance for (5.20), when $a(x) \sim x^{K_a}$ and $b(x) = x^{K_b}$ [68, 69] or $a(x) = x^{K_a}$, $b(x) = x^{K_b}$ [115], provided that $K_a + K_b \leq 2$, excluding the case $K_a = K_b = 1$. Moreover, if u_0 is only of class $L^2(0, 1)$, the solution is not sufficiently regular to verify the additional condition at x_0 established in [79, Lemma 2.11], and this procedure cannot be pursued. Hence, for these situations we can not deduce null controllability for (5.1) by known results.

5.2   The Non Divergence Form

The present section is devoted to give a full analysis of (5.1) in non divergence form. More precisely, we consider the problem

$$
\begin{cases}
u_t - a(x)u_{xx} - \dfrac{\lambda}{b(x)}u = f(t,x)\chi_\omega(x), & (t,x) \in Q_T, \\
u(t,0) = u(t,1) = 0, & t \in (0,T), \\
u(0,x) = u_0(x), & x \in (0,1),
\end{cases}
\tag{5.22}
$$

where $u_0 \in L^2_{\frac{1}{a}}(0,1)$, $f \in L^2_{\frac{1}{a}}(Q_T)$ and ω satisfies (2.4). Here $L^2_{\frac{1}{a}}(0,1)$ is the Hilbert space introduced in Sect. 2.1.2.

In order to study well posedness of problem (5.22) and in view of Lemmas 1.8 and 1.9 we shall use the space $\mathcal{K}$ defined in (1.38).

From now on, we make the following assumptions on a, b and λ:

Hypothesis 5.3 1. Hypothesis 1.3 holds;

2. $\lambda \neq 0$ and $\lambda < \dfrac{1}{C^*}$.   □

Again the assumption $\lambda \neq 0$ is not restrictive since the case $\lambda = 0$ is considered in [20, 78] (see also Chap. 2).

Using the lemmas given in Sect. 1.2 one can prove the next inequality, which is crucial to prove well posedness.

Proposition 5.2 (Proposition 3.1 [71]) *Assume Hypothesis 5.3. Then there exists $\Lambda > 0$ such that for all $u \in \mathcal{K}$*

$$
\int_0^1 (u'(x))^2 dx - \lambda \int_0^1 \frac{u^2(x)}{a(x)b(x)}dx \geq \Lambda \|u\|^2_{\mathcal{K}}.
$$

Finally, we introduce the Hilbert space

$$
\mathcal{H}^2_{\frac{1}{a},\frac{1}{b},x_0}(0,1) := \left\{ u \in \mathcal{H}^1_{\frac{1}{a},x_0}(0,1) \mid u' \in H^1(0,1) \text{ and } \mathcal{A}_\lambda u \in L^2_{\frac{1}{a}}(0,1) \right\}
$$

and consider

$$
D(\mathcal{A}_\lambda) = \mathcal{H}^2_{\frac{1}{a},\frac{1}{b},x_0}(0,1).
$$

Observe that if $u \in D(\mathcal{A}_\lambda)$, then $\dfrac{u}{b}$ and $\dfrac{u}{\sqrt{b}} \in L^2_{\frac{1}{a},x_0}(0,1)$, so that $u \in \mathcal{K}_{a,b,x_0}(0,1)$ and (1.37) holds if Hypothesis 1.3 is satisfied.

Thanks to Proposition 5.2 and the Green formula [71, Lemma 3.1], one can prove that (5.22) is well posed:

Theorem 5.5 (Theorem 3.1 [71]) *Assume Hypothesis 5.3. For all $f \in L^2_{\frac{1}{a}}(Q_T)$ and $u_0 \in L^2_{\frac{1}{a}}(0,1)$, there exists a unique weak solution u of (5.22). In particular, the* operator $(A_\lambda, D(A_\lambda))$ is non positive and self-adjoint in $L^2_{\frac{1}{a}}(0, 1)$. Moreover, if $u_0 \in D(A_\lambda)$ and $f \in W^{1,1}(0, T; L^2_{\frac{1}{a}}(0, 1))$, then

$$u \in C^1(0, T; L^2_{\frac{1}{a}}(0, 1)) \cap C([0, T]; D(A_\lambda)).$$

In order to study null controllability for (5.22), we consider, as usual, the non homogeneous adjoint problem

$$\begin{cases} v_t + a(x)v_{xx} + \dfrac{\lambda}{b(x)}v = h, & (t, x) \in Q_T, \\ v(t, 0) = v(t, 1) = 0, & t \in (0, T), \\ v(T, x) = v_T(x) \in L^2_{\frac{1}{a}}(0, 1). \end{cases} \qquad (5.23)$$

As usual, to prove Carleman estimates the final datum $v_T(x)$ is irrelevant, so we can omit it.

In the following we make the next assumptions:

Hypothesis 5.4

1. Hypothesis 5.3 is satisfied;
2. $\dfrac{(x - x_0)a'(x)}{a(x)} \in W^{1,\infty}(0, 1)$;
3. if $K_a \geq \dfrac{1}{2}$, there exists $\theta \in (0, K_a]$ such that the function $x \mapsto \dfrac{a}{|x - x_0|^\theta}$ is non-increasing on the left and nondecreasing on the right of $x = x_0$;
4. if $\lambda < 0$, then $(x - x_0)b'(x) \geq 0$ in $[0, 1]$. $\qquad\qquad\square$

Again, let us introduce the function $\varphi := \Theta\psi$, where Θ and ψ are as in (2.9) and (2.34), respectively, and define the space

$$\mathcal{V} := H^1(0, T; \mathcal{K}) \cap L^2(0, T; \mathcal{H}^2_{\frac{1}{a}, \frac{1}{b}, x_0}(0, 1)). \qquad (5.24)$$

The following Carleman estimate holds.

Theorem 5.6 (Theorem 4.1 [71]) *Assume Hypothesis 5.4. Then there exist two positive constants C and s_0 (depending on λ) such that every solution v of (5.23) in $\mathcal{V}$ satisfies, for all $s \geq s_0$,*

$$\int_0^T \int_0^1 \left(s\Theta v_x^2 + s^3\Theta^3 \left(\frac{x - x_0}{a}\right)^2 v^2\right) e^{2s\varphi}dxdt$$

$$\leq C\left(\int_0^T \int_0^1 h^2 \frac{e^{2s\varphi}}{a}dxdt + s\int_0^T \Theta\left[(x - x_0)v_x^2\right]_{x=0}^{x=1}dt\right). \qquad (5.25)$$

As usual, one can obtain the observability inequality for the homogeneous adjoint problem as a consequence of the Carleman estimate (5.25) (we refer to [71] for the proof of (1.45) under different and weaker assumptions). Hence, by Theorem 1.3, one obtains that

problem (5.22) is null controllable.

We conclude observing that the technique used in the proof of Theorem 5.4 can be applied also in the non divergence case. In this way, we obtain null controllability under weaker assumptions. Indeed, by using Carleman estimates, we need to require Hypothesis 5.4: for example, if $\lambda < 0$ then one has to ask the additional condition $(x - x_0)b'(x) \geq 0$ in $[0, 1]$. On the other hand, using the technique used in the proof of Theorem 5.4, one has to require only the conditions for the existence theorem (see Theorem 5.5 and Hypothesis 5.3). Indeed, proceeding as in the proof of Theorem 5.4, the control function f of (5.22) is given by

$$f = \phi_1 f_1 \chi_{(\zeta,x_0-r)} + \phi_2 f_2 \chi_{\cap(x_0+r,\xi)} - \frac{1}{T}\phi_0 u_3 - 2\phi_1' au_{1,x} - \phi_1'' au_1 - 2\phi_2' au_{2,x}$$
$$- \phi_2'' au_2 - \phi_0'\frac{T-t}{T}au_{3,x} - a\frac{T-t}{T}\left(\phi_0' u_3\right)_x,$$

if ω satisfies (2.29) or

$$f = \varphi_1 f_4 \chi_{(\zeta_1,\xi_1)} + \varphi_2 f_5 \chi_{(\zeta_2,\xi_2)} - \frac{1}{T}\varphi_0 u_3 - 2\varphi_1' au_{4,x} - \varphi_1'' au_4 - 2\phi_2' au_{5,x}$$
$$- \varphi_2'' au_5 - \varphi_0'\frac{T-t}{T}au_{3,x} - a\frac{T-t}{T}\left(\varphi_0' u_3\right)_x,$$

if ω satisfies (2.30). In every case f belongs to the space $L^2_{\frac{1}{a}}(Q_T)$ as required. Hence, the next theorem holds:

Theorem 5.7 (Theorem 2.6 [80]) *Assume Hypotheses 2.4 and 5.3. Then, given $u_0 \in L^2_{\frac{1}{a}}(0, 1)$, there exists $f \in L^2_{\frac{1}{a}}(Q_T)$ such that the solution u of (5.22) satisfies*

$$u(T, x) = 0 \text{ for every } x \in [0, 1].$$

Moreover

$$\int_0^T \int_0^1 \frac{f^2}{a} dx dt \leq C \int_0^1 \frac{u_0^2}{a} dx, \tag{5.26}$$

for some universal positive constant C.

We underline that Hypothesis 5.3 is just an assumption ensuring that problem (5.22) is well posed.

5.3 Further Developments

Also in this case, we conclude by giving a non exhaustive short list of possible directions of research.

Systems with interior degeneracy and singularity: controllability and Carleman estimates are considered also for systems in which the equations are governed by degenerate and singular operators having their degeneracy at an interior point. For instance, see [21, 92, 107]. See also [4] for an inverse source problem for a single equation.

Problems with Neumann boundary conditions: In all the examples considered in these notes, we have always treated problems with Dirichlet boundary conditions. This means that the underlying space is H_0^1 or some of its weighted versions. On the other hand, if Neumann boundary conditions are in force, the reference space is H^1 (or its weighted versions); it is clear that Hardy inequalities like

$$\left\| \frac{u}{x^\alpha} \right\|_{L^2} \leq C \|x^\beta u'\|_{L^2} \tag{5.27}$$

for some $\alpha, \beta > 0$ cannot hold because of the presence of nonzero constants. However, one can prove suitable modified versions of (5.27) with additional terms on the right-hand-side of the inequality (for instance, see [71]), and thus related controllability results can be found in [20, 71, 81–83].

References

1. Alabau-Boussouira, F., Cannarsa, P., Fragnelli, G.: Carleman estimates for degenerate parabolic operators with applications to null controllability. J. Evol. Equ. **6**, 161–204 (2006)
2. Alabau-Boussouira, F., Cannarsa, P., Leugering, G.: Control and stabilization of degenerate wave equations. SIAM J. Control Optim. **55**, 2052–2087 (2017)
3. Ammar-Khodja, F., Benabdallah, A., Gonzlez-Burgos, M., de Teresa, L.: Recent results on the controllability of linear coupled parabolic problems: a survey. Math. Control Relat. Fields **1**, 267–306 (2011)
4. Atifi, K., Boutaayamou, I., Sidi, H.O., Salhi, J.: An inverse source problem for singular parabolic equations with interior degeneracy. Abstr. Appl. Anal. Art 2067304 (2018), 16 pp
5. Baras, P., Goldstein, J.: Remarks on the inverse square potential in quantum mechanics. Differ. Equ. N.-Holl. Math. Stud. **92**, 31–35 (1984)
6. Baras, P., Goldstein, J.: The heat equation with a singular potential. Trans. Am. Math. Soc. **284**, 121–139 (1984)
7. Barbu, V., Favini, A., Romanelli, S.: Degenerate evolution equations and regularity of their associated semigroups. Funkcial. Ekvac **39**, 421–448 (1996)
8. Bebernes, J., Eberly, D.: Mathematical Problems from Combustion Theory. Mathematical Sciences, vol. 83. Springer, New York (1989)
9. Beauchard, K., Cannarsa, P., Guglielmi, R.: Null controllability of Grushin-type operators in dimension two. J. Eur. Math. Soc. **16**, 67–101 (2014)
10. Bellassoued, M., Yamamoto, M.: Carleman Estimates and Applications to Inverse Problems for Hyperbolic Systems. Springer Monographs in Mathematics (2017)
11. Ben Hassi, E.M.A., Fadili, M., Maniar, L.: Controllability of a system of degenerate parabolic equations with non-diagonalizable diffusion matrix. Math. Control Relat. Fields **10**, 623–642 (2020)
12. Benaissa, A., Aichi, C.: Energy decay for a degenerate wave equation under fractional derivative controls. Filomat **32**, 6045–6072 (2018)
13. Bensoussan, A., Da Prato, G., Delfour, M.C., Mitter, S.K.: Representation and Control of Infinite Dimensional Systems, vol. 1. Birkhäuser, Boston (1992)
14. Berger, M., Gauduchon, P., Mazet, E.: Le spectre d'une variété Riemannienne. Springer, Berlin (1971)
15. Bhandari, K., Boyer, F.: Boundary null-controllability of coupled parabolic systems with Robin conditions. Evol. Equ. Control Theory **10**, 61–102 (2021)

© The Author(s), under exclusive license to Springer Nature Switzerland AG 2021

G. Fragnelli and D. Mugnai, *Control of Degenerate and Singular Parabolic Equations*,
SpringerBriefs in Mathematics, https://doi.org/10.1007/978-3-030-69349-7

16. Biccari, U., Hernandez-Santamaria, V., Vancostenoble, J.: Existence and cost of boundary controls for a degenerate/singular parabolic equation (2020). arXiv:2001.11403
17. Boutaayamou, I., Echarroudi, Y.: Null controllability of a population dynamics with interior degeneracy. Electron. J. Differ. Equ. **2017**(131), 1–21 (2017)
18. Boutaayamou, I., Fragnelli, G.: A degenerate population system: Carleman estimates and controllability. Nonlinear Anal. **195**, 111742 (2020)
19. Boutaayamou, I., Fragnelli, G., Maniar, L.: Lipschitz stability for linear cascade parabolic systems with interior degeneracy. Electron. J. Differ. Equ. **2014**, 1–26 (2014)
20. Boutaayamou, I., Fragnelli, G., Maniar, L.: Carleman estimates for parabolic equations with interior degeneracy and Neumann boundary conditions. J. Anal. Math. **135**, 1–35 (2018)
21. Boutaayamou, I., Salhi, J.: Null controllability for linear parabolic cascade systems with interior degeneracy. Electron. J. Differ. Equ. **2016**(305), 1–22 (2016)
22. Brezis, H.: Functional Analysis, Sobolev Spaces and Partial Differential Equations. Springer Science+Business Media, LLC, New York (2011)
23. Brezis, H., Vazquez, J.L.: Blow-up solutions of some nonlinear elliptic equations. Rev. Mat. Complut. **10**, 443–469 (1997)
24. Cabré, X., Martel, Y.: Existence versus explosion instantanée pour des équations de la chaleur linéaires avec potentiel singulier. C. R. Math. Acad. Sci. Paris **329**, 973–978 (1999)
25. Campiti, M., Metafune, G., Pallara, D.: Degenerate self-adjoint evolution equations on the unit interval. Semigroup Forum **57**, 1–36 (1998)
26. Cannarsa, P., Ferretti, R., Martinez, P.: Null controllability for parabolic operators with interior degeneracy and one-sided control. SIAM J. Control Optim. **57**, 900–924 (2019)
27. Cannarsa, P., Floridia, G.: Approximate multiplicative controllability for degenerate parabolic problems with Robin boundary conditions. Commun. Appl. Ind. Math. **2**(2), Article ID 376 (2011)
28. Cannarsa, P., Fragnelli, G.: Null controllability of semilinear degenerate parabolic equations in bounded domains. Electron. J. Differ. Equ. **2006**(136), 1–20 (2006)
29. Cannarsa, P., Fragnelli, G., Rocchetti, D.: Controllability results for a class of one-dimensional degenerate parabolic problems in nondivergence form. J. Evol. Equ. **8**, 583–616 (2008)
30. Cannarsa, P., Fragnelli, G., Rocchetti, D.: Null controllability of degenerate parabolic operators with drift. Netw. Heterog. Media **2**, 693–713 (2007)
31. Cannarsa, P., Fragnelli, G., Vancostenoble, J.: Linear degenerate parabolic equations in bounded domains: controllability and observability. IFIP Int. Fed. Inf. Process. **202** (Springer, New York, 2006), 163–173
32. Cannarsa, P., Fragnelli, G., Vancostenoble, J.: Regional controllability of semilinear degenerate parabolic equations in bounded domains. J. Math. Anal. Appl. **320**, 804–818 (2006)
33. Cannarsa, P., Fragnelli, G., Vancostenoble, J.: Regional controllability of semilinear parabolic equations in unbounded domains. In: Proceedings of the Sixth Portuguese Conference on Automatic Control, Faro, Portugal, 7–9 June (CONTROLO 2004)
34. Cannarsa, P., Martinez, P., Vancostenoble, J.: Carleman estimates for a class of degenerate parabolic operators. SIAM J. Control Optim. **47**, 1–19 (2008)
35. Cannarsa, P., Martinez, P., Vancostenoble, J.: Global Carleman estimates for degenerate parabolic operators with applications. Mem. Am. Math. Soc. **239** (2016), ix+209 pp
36. Cannarsa, P., Martinez, P., Vancostenoble, J.: Null controllability of degenerate heat equations. Adv. Differ. Equ. **10**, 153–190 (2005)
37. Cannarsa, P., Martinez, P., Vancostenoble, J.: Persistent regional controllability for a class of degenerate parabolic equations. Commun. Pure Appl. Anal. **3**, 607–635 (2004)
38. Cannarsa, P., Martinez, P., Vancostenoble, J.: The cost of controlling weakly degenerate parabolic equations by boundary controls. Math. Control Relat. Fields **7**, 171–211 (2017)
39. Cannarsa, P., Tort, J., Yamamoto, M.: Unique continuation and approximate controllability for a degenerate parabolic equation. Appl. Anal. **91**, 1409–1425 (2012)
40. Carleman, T.: Sur un problème d'unicité pour les systèmes d'équations aux dérivées partielles à deux variables indépendantes. Ark. Mat. Astr. Fys. **26**, 1–9 (1939)

41. Cazacu, C.: Controllability of the heat equation with an inverse-square potential localized on the boundary. SIAM J. Control Optim. **52**, 2055–2089 (2014)
42. Coron, J.-M.: Control and Nonlinearity. Mathematical Surveys and Monographs, vol. 136. American Mathematical Society, Providence (2007)
43. Coron, J.-M.: Global asymptotic stabilization for controllable systems without drift. Math. Control Signals Syst. **5**, 227–232 (1992)
44. Dautray, R., Lions, J.-L.: Analyse Mathématique et Calcul Numérique pour les Sciences et les Tehniques. Collection du Commissatiat à l'Énergie Atomique: Séries Scientifique, vol. 2. Masson, Paris (1985)
45. Dautray, R., Lions, J.-L.: Mathematical Analysis and Numerical Methods for Science and Technology. Functional and Variational Methods, vol. 2. Springer, Berlin (1988)
46. Davies, E.B.: Spectral Theory and Differential Operators. Cambridge Stud. Adv. Math., vol. 42. Cambridge University Press, Cambridge (1995)
47. De Castro, A.S.: Bound states of the Dirac equation for a class of effective quadratic plus inversely quadratic potentials. Ann. Phys. **311**, 170–181 (2004)
48. Dold, J.W., Galaktionov, V.A., Lacey, A.A., Vazquez, J.L.: Rate of approach to a singular steady state in quasilinear reaction-diffusion equations. Ann. Sc. Norm. Super. Pisa Cl. Sci. **26**, 663–687 (1998)
49. Dolecki, S., Russell, D.L.: A general theory of observation and control. SIAM J. Control Optim. **15**, 185–220 (1977)
50. Du, R., Xu, F.: On the boundary controllability of a semilinear degenerate system with the convection term. Appl. Math. Comput. **303**, 113–127 (2017)
51. Engel, K.J., Nagel, R.: One-Parameter Semigroups for Linear Evolution Equations. Springer, New York (2000)
52. Epstein, C.L., Mazzeo, R.: Degenerate Diffusion Operators Arising in Population Biology. Ann. Math. Stud. (2013)
53. Ervedoza, S.: Control and stabilization properties for a singular heat equation with an inverse-square potential. Commun. Partial Differ. Equ. **33**, 1996–2019 (2008)
54. Escauriaza, L., Seregin, G., Sverak, V.: Backward uniqueness for parabolic equations. Arch. Ration. Mech. Anal. **169**, 147–157 (2003)
55. Escauriaza, L., Seregin, G., Sverak, V.: Backward uniqueness for the heat operator in half-space. St. Petersburg Math. J. **15**, 139–148 (2004)
56. Fabre, C., Puel, J.P., Zuazua, E.: Approximate controllability of the semilinear heat equation. Proc. R. Soc. Edinb. **125A**, 31–61 (1995)
57. Fall, M.M.: On the Hardy-Poincaré inequality with boundary singularities. Commun. Contemp. Math. **14**(3), 1250019, 13 (2012)
58. Fall, M.M., Musina, R.: Hardy-Poincaré inequalities with boundary singularities. Proc. R. Soc. Edinb. A **142**, 769–786 (2012)
59. Fardigola, L.V.: Transformation operators in the problems of controllability for the degenerate wave equation with variable coefficients. Ukr. Math. J. **70**, 1300–1318 (2019)
60. Fattorini, H.O.: On complete controllability of linear systems. J. Differ. Equ. **3**, 391–402 (1967)
61. Fattorini, H.O., Russell, D.L.: Exact controllability theorems for linear parabolic equations in one space dimension. Arch. Ration. Mech. Anal. **43**, 272–292 (1971)
62. Fernández-Cara, E.: Null controllability of the semilinear heat equation. ESAIM: Control Optim. Calc. Var. **2**, 87–103 (1997)
63. Fernandez-Cara, E., Gonzalez-Burgos, M., Guerrero, S., Puel, J.P.: Exact controllability to the trajectories of the heat equation with Fourier boundary conditions: the semilinear case. ESAIM Control Optim. Calc. Var. **12**, 466–483 (2006)
64. Fernandez-Cara, E., Gonzalez-Burgos, M., Guerrero, S., Puel, J.P.: Null controllability of the heat equation with boundary Fourier conditions: the linear case. ESAIM Control Optim. Calc. Var. **12**, 442–465 (2006)
65. Fernández-Cara, E., Guerrero, S.: Global Carleman inequalities for parabolic systems and applications to controllability. SIAM J. Control Optim. **45**, 1395–1446 (2006)

66. Fernández-Cara, E., Zuazua, E.: Null and approximate controllability for weakly blowing up semilinear heat equations. Ann. Inst. H. Poincaré Anal. Non Linéaire **17**, 583–616 (2000)
67. Fernández-Cara, E., Zuazua, E.: The cost of approximate controllability for heat equations: the linear case. Adv. Differ. Equ. **5**, 465–514 (2000)
68. Fotouhi, M., Salimi, L.: Controllability results for a class of one dimensional degenerate/singular parabolic equations. Commun. Pure Appl. Anal. **12**, 1415–1430 (2013)
69. Fotouhi, M., Salimi, L.: Null controllability of degenerate/singular parabolic equations. J. Dyn. Control Syst. **18**, 573–602 (2012)
70. Fragnelli, G.: Null controllability of degenerate parabolic equations in non divergence form via Carleman estimates. Discret. Contin. Dyn. Syst. Ser. S **6**, 687–701 (2013)
71. Fragnelli, G.: Interior degenerate/singular parabolic equations in nondivergence form: well-posedness and Carleman estimates. J. Differ. Equ. **260**, 1314–1371 (2016)
72. Fragnelli, G.: Null controllability for a degenerate population model in divergence form via Carleman estimates. Adv. Nonlinear Anal. **9**, 1102–1129 (2020)
73. Fragnelli, G.: Controllability for a population equation with interior degeneracy. Pure Appl. Funct. Anal. **4**, 803–824 (2019)
74. Fragnelli, G.: Carleman estimates and null controllability for a degenerate population model. J. Math. Pures Appl. **115**(9), 74–126 (2018)
75. Fragnelli, G., Marinoschi, G., Mininni, R.M., Romanelli, S.: A control approach for an identification problem associated to a strongly degenerate parabolic system with interior degeneracy. In: Favini, A., Fragnelli, G., Mininni, R.M. (eds.) New-Prospects in Direct, Inverse and Control Problems for Evolution Equation. Springer INdAM Series, vol. 10, pp. 121–139 (2014)
76. Fragnelli, G., Marinoschi, G., Mininni, R.M., Romanelli, S.: Identification of a diffusion coefficient in strongly degenerate parabolic equations with interior degeneracy. J. Evol. Equ. **15**, 27–51 (2015)
77. Fragnelli, G., Mugnai, D.: Carleman estimates and observability inequalities for parabolic equations with interior degeneracy. Adv. Nonlinear Anal. **2**, 339–378 (2013)
78. Fragnelli, G., Mugnai, D.: Carleman estimates, observability inequalities and null controllability for interior degenerate non smooth parabolic equations. Mem. Am. Math. Soc. **242** (2016), v+84 pp
79. Fragnelli, G., Mugnai, D.: Carleman estimates for singular parabolic equations with interior degeneracy and non smooth coefficients. Adv. Nonlinear Anal. **6**, 61–84 (2017)
80. Fragnelli, G., Mugnai, D.: Controllability of strongly degenerate parabolic problems with strongly singular potentials. Electron. J. Qual. Theory Differ. Equ. **50**, 1–11 (2018)
81. Fragnelli, G., Mugnai, D.: Controllability of degenerate and singular parabolic problems: the double strong case with Neumann boundary conditions. Opusc. Math. **39**, 207–225 (2019)
82. Fragnelli, G., Mugnai, D.: Singular parabolic equations with interior degeneracy and non smooth coefficients: the Neumann case. Discret. Contin. Dyn. Syst.-S **13**, 1495–1511 (2020)
83. Fragnelli, G., Mugnai, D.: Corrigendum and improvements to Carleman estimates, observability inequalities and null controllability for interior degenerate non smooth parabolic equations, and its consequences. To appear in Mem. Am. Math. Soc. (2021)
84. Fragnelli, G., Ruiz Goldstein, G., Goldstein, J.A., Romanelli, S.: Generators with interior degeneracy on spaces of L^2 type. Electron. J. Differ. Equ. **2012**, 1–30 (2012)
85. Fragnelli, G., Yamamoto, M.: Carleman estimates and controllability for a degenerate structured population model. Appl. Math. Optim. https://doi.org/10.1007/s00245-020-09669-0
86. Fu, X., Lü, Q., Zhang, X.: Carleman Estimates for Second Order Partial Differential Operators and Applications. Springer, Berlin (2019)
87. Fursikov, A.V., Imanuvilov, O.Yu.: Controllability of Evolution Equations. Lecture Notes Series, Research Institute of Mathematics, Global Analysis Research Center, Seoul National University, vol. 34 (1996)
88. Galaktionov, V., Vazquez, J.L.: Continuation of blow-up solutions of nonlinear heat equations in several space dimensions. Commun. Pure Appl. Math. **50**, 1–67 (1997)

89. Gueye, M.: Exact boundary controllability of 1-D parabolic and hyperbolic degenerate equations. SIAM J. Control Optim. **52**, 2037–2054 (2014)
90. Gueye, M., Lissy, P.: Singular optimal control of a 1-D parabolic-hyperbolic degenerate equation. ESAIM Control Optim. Calc. Var. **22**, 1184–1203 (2016)
91. Hagan, P., Woodward, D.: Equivalent Black volatilities. Appl. Math. Financ. **6**, 147–159 (1999)
92. Hajjaj, A., Maniar, L., Salhi, J.: Carleman estimates and null controllability of degenerate/singular parabolic systems. Electron. J. Differ. Equ. **2016**(292), 1–25 (2016)
93. Jiang, F., Luan, Y., Li, G.: Asymptotic stability and blow-up of solutions for an edge-degenerate wave equation with singular potentials and several nonlinear source terms of different sign. Electron. J. Differ. Equ. **2018**(18), 1–27 (2018)
94. Karachalios, N.I., Zographopoulos, N.B.: On the dynamics of a degenerate parabolic equation: global bifurcation of stationary states and convergence. Calc. Var. Partial Differ. Equ. **25**, 361–393 (2006)
95. Lebeau, G., Robbiano, L.: Contrôle exact de l'équation de la chaleur. Comm. P.D.E. **20**, 335–356 (1995)
96. Lions, J.L.: Contrôlabilité exacte, perturbations et stabilisation de systèmes distribués. 1, Recherches en Mathématiques Appliquées, 8. Paris etc.: Masson. x (1988), 538 p
97. Lions, J.L.: Remarks on approximate controllability. J. Anal. Math. **59**, 103–116 (1992)
98. Lions, J.-L., Magenes, E.: Non-homogeneous Boundary Value Problems and Applications. Grundlehren der mathematischen Wissenschaften, vol. 1, p. 181. Springer, New York (1972)
99. Lions, J.-L., Magenes, E.: Non-homogeneous Boundary Value Problems and Applications. Grundlehren der mathematischen Wissenschaften, vol. 2, p. 181. Springer, New York (1972)
100. Martin, P., Rosier, L., Rouchon, P.: Null controllability of one-dimensional parabolic equations by the flatness approach. SIAM J. Control Optim. **54**, 198–220 (2016)
101. Martinez, P., Vancostenoble, J.: Carleman estimates for one-dimensional degenerate heat equations. J. Evol. Equ. **6**, 325–362 (2006)
102. Micu, S., Zuazua, E.: On the lack of null controllability of the heat equation on the half-line. Trans. Am. Math. Soc. **353**, 1635–1659 (2001)
103. Miller, L.: The control transmutation method and the cost of fast controls. SIAM J. Control Optim. **45**, 762–772 (2006)
104. Mizel, V.J., Seidman, T.I.: Observation and prediction for the heat equation. J. Math. Anal. Appl. **28**, 303–312 (1969)
105. Muckenhoupt, B.: Hardy's inequality with weights. Collection of articles honoring the completion by Antoni Zygmund of 50 years of scientific activity, I. Studia Math. **44**, 31–38 (1972)
106. Renardy, M., Rogers, R.C.: An Introduction to Partial Differential Equations. Texts in Applied Mathematics, vol. 13, 2nd edn. Springer, New York (2004)
107. Salhi, J.: Null controllability for a coupled system of degenerate/singular parabolic equations in nondivergence form. Electron. J. Qual. Theory Differ. Equ. **31**, 1–28 (2018)
108. Seidman, T.I., Avdonin, S.A., Ivanov, S.A.: The "window problem" for series of complex exponentials. J. Fourier Anal. Appl. **6**, 233–254 (2000)
109. Showalter, R.E.: Hilbert space methods for partial differential equations. Electronic reprint of the 1977 original. Electron. J. Differ. Equ. Monogr. 1. San Marcos, TX (1994)
110. Strauss, W.A.: Partial Differential Equations. An Introduction. Wiley, New York (1992)
111. Tataru, D.: A-priori estimates of Carlemans type in domains with boundary. J. Math. Pures Appl. **73**(9), 355–387 (1994)
112. Tataru, D.: Carleman estimates, unique continuation and controllability for anisotropic PDE's. Contemp. Math. **209**, 267–279 (1997)
113. Tenenbaum, G., Tucsnak, M.: New blow-up rates for fast controls of Schrödinger and heat equations. J. Differ. Equ. **243**, 70–100 (2007)
114. Vancostenoble, J.: Lipschitz stability in inverse source problems for singular parabolic equations. Commun. Partial Differ. Equ. **36**, 1287–1317 (2011)
115. Vancostenoble, J.: Improved Hardy-Poincaré inequalities and sharp Carleman estimates for degenerate/singular parabolic problems. Discret. Contin. Dyn. Syst. Ser. S **4**, 761–790 (2011)

116. Vancostenoble, J., Zuazua, E.: Hardy inequalities, observability, and control for the wave and Schrödinger equations with singular potentials. SIAM J. Math. Anal. **41**, 1508–1532 (2009)
117. Vancostenoble, J., Zuazua, E.: Null controllability for the heat equation with singular inverse-square potentials. J. Funct. Anal. **254**, 1864–1902 (2008)
118. Vázquez, J.L., Zuazua, E.: The Hardy inequality and the asymptotic behaviour of the heat equation with an inverse-square potential. J. Funct. Anal. **173**, 103–153 (2000)
119. Wang, C., Du, R.: Carleman estimates and null controllability for a class of degenerate parabolic equations with convection terms. SIAM J. Control Optim. **52**, 1457–1480 (2014)
120. Xu, F., Zhou, Q., Nie, Y.: Null controllability of a semilinear degenerate parabolic equation with a gradient term. Bound. Value Probl. **2020**, 55 (2020)
121. Xu, J., Wang, C., Nie, Y.: Carleman estimate and null controllability of a cascade degenerate parabolic system with general convection terms. Electron. J. Differ. Equ. **2018**(195), 1–20 (2018)
122. Zhang, X.: A remark on null controllability of the heat equation. SIAM J. Control Optim. **40**, 39–53 (2001)
123. Zhang, M., Gao, H.: Interior controllability of semi-linear degenerate wave equations. J. Math. Anal. Appl. **457**, 10–22 (2018)
124. Zhang, M., Gao, H.: Null controllability of some degenerate wave equations. J. Syst. Sci. Complex. **30**, 1027–1041 (2017)
125. Zhang, M., Gao, H.: Persistent regional null controllability of some degenerate wave equations. Math. Methods Appl. Sci. **40**, 5821–5830 (2017)
126. Zuazua, E.: Approximate controllability for the semilinear heat equation with globally Lipschitz nonlinearities. Control Cybern. **28**, 665–683 (1999)

Index

© The Author(s), under exclusive license to Springer Nature Switzerland AG 2021   105
G. Fragnelli and D. Mugnai, *Control of Degenerate and Singular Parabolic Equations*,
SpringerBriefs in Mathematics, https://doi.org/10.1007/978-3-030-69349-7

Printed in the United States
by Baker & Taylor Publisher Services